CITIZENS OF THE SEA

▲ This leafy sea dragon (*Phycodurus eques*), found off the coast of southern Australia, looks conspicuous here but is actually very hard to spot when surrounded by seaweed.

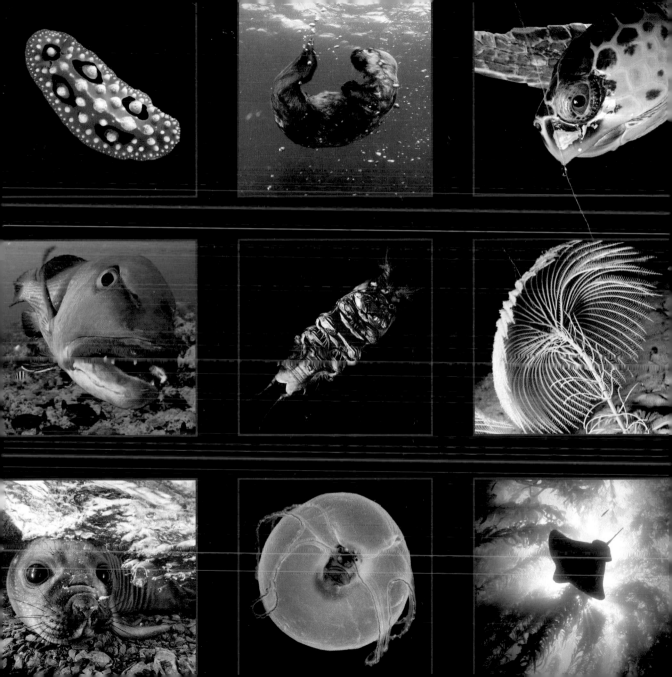

CITIZENS OF THE
SEA

Wondrous Creatures from the Census of Marine Life

CONTENTS

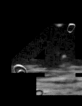

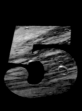

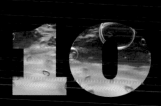
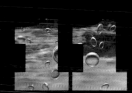
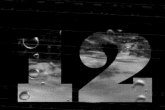
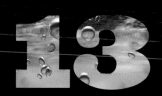
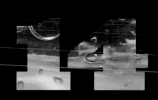

FOREWORD by Enric Sala

IT IS A CURIOUS AND DISTRESSING FACT THAT, IN SPITE OF ALL WE KNOW about life on Earth, so many basic questions still remain, especially when it comes to life in the ocean. What lies underneath the large ocean that covers more than 70 percent of the planet's surface? We know that it's full of salty water and that life can be found in every drop of it—actually, tens of thousands of bacteria in every drop. And, in a place like the western Mediterranean, we can discover as many as 100 different species of plants in an area the size of a dinner plate. But, when it comes to guessing how much life there is in the entire ocean, we are unsure by a factor of 20. Experts estimate that there may be between 500,000 and 10 million different marine species—some difference!

This is why the Census of Marine Life was launched. The Census is an international effort with the goal of answering three basic questions: What lived, lives, and will live in the world's ocean? Over the past decade, 2,000 scientists searched the ocean from Pole to Pole, from top to bottom, for everything from microbes to whales. The goal was not just to estimate the number of species but also to figure out how much their numbers have changed over time and what the future of marine life might be.

The Census has been a journey of discovery. There is no agreed-upon method for making a census of marine life. So the Census used an eclectic array of tools—from scuba diving to satellites—and human ingenuity to construct an impressive collage of information, from the discovery of new species of deep-sea creatures to the tracking of the migration of large ocean predators such as tuna. It's detective work—fascinating, challenging, and often exhilarating.

Take reef fish. How can we know how many there are on a particular coral reef? You swim along a 50-meter line, identifying, counting, and estimating the size of every fish within two meters on each side. This is harder than it sounds. Imagine yourself swimming against a current, with schools of hundreds of small fishes moving frantically from one side of the line to the other while other fish dive down into the reef to hide. To complicate things, males and females are colored differently in some species, just like some birds. In addition, large fishes tend to swim away as fast as fishly possible, which makes it very difficult for one to estimate how big they are. Now multiply this times 300 species or more.

For the rest of the ocean, mostly beyond what our human bodies can tolerate, the quest is even more daunting. Exploring the deep sea is similar to space exploration: We need sophisticated submersibles to isolate us from the hostile environment—dark, cold, and under intense pressure. We can catch only glimpses of a world where we are welcome only as temporary guests.

If counting the number of species is difficult, understanding what they do is even more complicated. Why do bluefin tuna cross the Atlantic to reproduce? Why do leatherback turtles swim over many thousands of kilometers to eat ubiquitous jellyfish? What is the role of the superabundant marine viruses? Some might ask: Why do we need to know? The answer is simple. The diversity of life in the sea makes our planet richer and sustains us, providing, for example, more than half of the oxygen we breathe.

Drawing on the many discoveries of the Census of Marine Life, Nancy Knowlton's book explores all the key issues—the diversity of ocean life and why it matters, how ocean creatures make a living, how we are changing life in the ocean, and what we can do to keep the ocean healthy. Nancy's deep knowledge of the sea and its citizens, gained by thousands of hours studying them underwater, is a gift to the reader. And, through her enjoyable and sometimes humorous prose, she allows readers to experience the many wonders of life in the sea.

Citizens of the Sea promises to take you on a unique journey. You will be inspired to see what's beneath the surface yourself and learn to care deeply about it.

ENRIC SALA *is Ocean Fellow at National Geographic Society in Washington, D.C., where he is leading an international marine conservation initiative. He is also ex-officio member of the Census of Marine Life Scientific Steering Committee, and researcher at Spain's National Research Council.*

▲ Spaghetti worms, like this species of *Loimia* from the Great Barrier Reef of Australia, are usually mostly hidden, except for their tentacles, which they spread out to trap the small particles that they feed on.

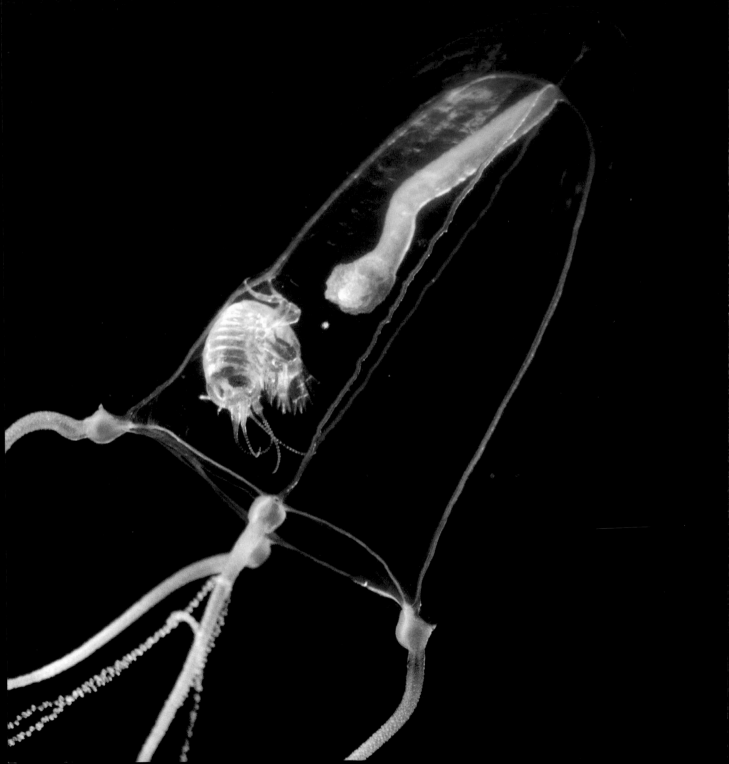

INTRODUCTION

T HE END OF THE FIRST CENSUS OF MARINE LIFE IN 2010 COINCIDES with the United Nations' International Year of Biodiversity and the 23rd United States Census. How appropriate, since these three large and superficially very different efforts have at a deep level much in common.

The ocean, at least beyond the surf zone, is out of sight and out of mind for most people. Sea creatures—even if familiar from trips to tide pools, aquariums, or restaurants—are mostly hard to relate to. A movie review of *Finding Nemo* ended tellingly by asking, "How expressive is a fish, until it hits a grill?"

Who and how many are the citizens of the sea? How and where do they live? What challenges do they face? Why should we care? I have tried to answer these and other questions in everyday language for a handful of ocean creatures, some studied by Census of Marine Life scientists and some not, chosen for their ability to instruct, amuse, and amaze. In the process I have learned a lot myself, despite my 35-plus years as an ocean scientist. So I hope this book speaks to anyone who might want to know more about ocean life—students and teachers, fishers and beachcombers, snorkelers and surfers, policymakers and the public they serve.

At the end of the ten-year Census of Marine Life, most ocean organisms remain nameless and their numbers unknown. This is not an admission of failure. The ocean is simply so vast that after a decade of hard work, we still have only snapshots, though sometimes detailed, of what the sea contains. But it is an important and impressive start. The stories in this book draw on discoveries of Census scientists and their colleagues past and present. Although not all of the images were taken by Census scientists, they were chosen because they depict the variety, beauty, weirdness, and wonder that characterizes life in the sea.

The sea today is in trouble. Its citizens have no vote in any national or international body, but they are suffering and need to be heard. Much has changed just in the few decades that I have spent on and under the sea, but it remains a wondrous and enriching place, and with care it can become even more so. That is why I wrote this book.

—Nancy Knowlton

◀ This shrimplike amphipod *(Hyperia medusarum)* lives as a parasite on jellyfish (like this *Sarsia* species) and other floating gelatinous organisms.

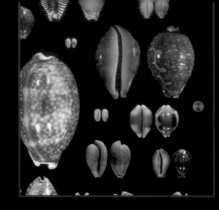

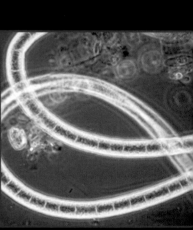

NAMES & NUMBERS

The Census of Marine Life, like censuses of people, is all about names and numbers. Most ocean organisms remain nameless, and even those with names can be hard to identify (though DNA barcodes help). The sea boasts the largest animal ever to live on Earth, and there are more microscopic creatures in the ocean than there are stars in the universe. Many ocean species are found on coral reefs, the rain forests of the sea.

WHAT'S THAT?

FEATURED: *Sarcastic Fringeheads, Yeti Crabs*

A S ANY PARENT OF A TODDLER KNOWS, NAMES ARE where language starts. Most animals and plants that live in the ocean are not commonly encountered and have no common names. But animals that we eat or fear or that appeal to amateur naturalists often do have names that use everyday nouns and adjectives—the queen conch, the tiger shark, the right whale.

Names were and are given for a reason—to convey something about the name holder, like the tiger shark with its stripes. Sometimes knowing a bit of history helps. The right whale, for example, got its name because it was the right whale to hunt—it was easy to catch and floated when dead. Occasionally names change, particularly if a change will bring in more cash. Hence the transformation of the "Patagonian toothfish" into the more edible-sounding "Chilean sea bass."

The origins of some names are more obscure. The sarcastic fringehead is a smallish, territorial, crevice-dwelling fish renowned for its huge mouth and fighting spirit (and willingness to latch onto the fingers of fishermen). Any intruder is tested in a mouth-wrestling match to determine who is the biggest fish on the block. As to the name, "fringehead" clearly refers to the ruff around its neck, but "sarcastic"? Yet the inexplicable is also sometimes unforgettable.

Scientists can be rather snobby about common names because there are no standards, and names vary from place to place. Fish-Base, the Internet source for all things piscine, lists more than 30 common names for the king salmon, which is scientifically known as *Oncorhynchus tshawytscha*—precise to be sure, but a mouthful. Yet even scientists know the power of an evocative name. Deep-sea explorers have hit bonanzas of late with names like the blobfish and the Dumbo octopus (now jumbo Dumbo). The recently discovered *Kiwa hirsuta* could have been dubbed the "hairy-armed crab," but the far more inspired choice of "Yeti crab" is surely what brought it Internet stardom (currently 6,590,000 Google hits!).

▲ The yeti crab (*Kiwa hirsuta*) was discovered at a depth of more than 7,000 feet. Unlike its fictional namesake, it thrives in the heat of hydrothermal vents.

FAST FACT:
You can search for common (or scientific) names at WoRMS—the World Register of Marine Species: *www.marinespecies.org*. This list will always be a work in progress—in the past ten years, tens of thousands of potential new species have been discovered, thanks to the Census of Marine Life.

◀ With its eyes rolled upward, a sarcastic fringehead fish (*Neoclinus blanchardi*) hiding in a squid's egg mass seems to live up to its name.

THE SCIENCE OF NAMES

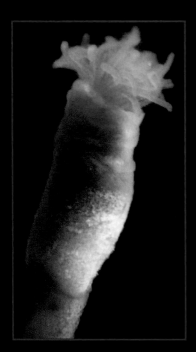

▲ Named for its "dreadlock" tentacles, the Bob Marley worm *(Bobmarleya gadensis)* lives inside a transparent tube near underwater mud volcanoes southwest of Spain.

FAST FACT:
Naming species can be more challenging than you might think (although rule books help). The hardest part is making sure no one has named the species before. Every species is epitomized by a single "type" specimen in a museum. If a species has been named more than once, by mistake, the oldest name usually wins out.

MORE THAN 270 YEARS AGO, THE SWEDISH BOTANIST Carl von Linné (in Latin, Linnaeus) created a system for naming and classifying organisms, which we still use. Two of its features stand out. First, it is a nested hierarchy, grouping ever larger assemblages of related organisms as one moves from species to genus to family to order to class to phylum to kingdom. Second, it assigns a unique, two-part name—genus and species (always italicized)—to each type of organism, as in *Homo sapiens* for us.

Consider the giant clam, *Tridacna gigas*, one of a surprisingly long list of animals and plants named by Linnaeus himself (although he called it *Chama gigas*). It belongs to the kingdom Animalia, the phylum Mollusca, the class Bivalvia, the order Euheterodonta, and the family Cardiidae. With some knowledge of classical languages, the characteristics of these groups can be guessed, although the original motivations for some have become obsolete. Animalia comes from "anima," meaning breath or soul. Mollusca confusingly means "soft," because slugs and cuttlefish (which are relatives of clams and snails) were grouped by Linnaeus with a hodgepodge of unrelated shell-less animals. Bivalvia again makes sense, coming from a word for the paired sides of folding doors. Euheterodonta refers to the unequal ("hetero") teeth ("donta") that hinge the shells. Heart-shaped clams are in the cockle family Cardiidae, as in "cardiac," or "to warm the cockles of my heart." Finally, *Tridacna*, from Greek, refers to something so large that three (big!) bites are needed to eat it, and *gigas* means giant. Today's scientists have many additional sources of nomenclatural inspiration—witness the worm with many tentacles named *Bobmarleya gadensis*.

Working more than a century before Darwin, Linnaeus was a creationist, not an evolutionist. He believed God created 312 genera (the plural of genus) and 4,000 species of animals, whereas today there are over 50,000 genera and many millions of species.

▶ On a Micronesian coral reef, this giant clam *(Tridacna gigas)* dwarfs its coral neighbors. Like corals, it gets energy from symbiotic algae in its tissues.

BARCODES

ABOUT 230,000 MARINE ORGANISMS HAVE NAMES, A fraction of the millions of species believed to live in the sea. At the rate new species are being described, it will take centuries to give them all names. But even if every species had a name, identification would not always be easy. The eggs and babies of different fish species often look alike, and even adults can be hard to distinguish once they are turned into filets. The solution? Barcodes. Not the string of tiny black lines scanned at the cash register, but the string of DNA letters in a gene (A, C, G, or T) that are, for certain genes, unique to a species.

Whales are an example of barcoding in action. Japan claims the right to harvest a limited number of whales of certain species for scientific research, the remains of which can then be sold. To check on the legality of what was on sale, two scientists bought 16 specimens at Tokyo markets for genetic analysis. By copying the DNA in their clandestine hotel laboratory and bringing back only the copies for analysis, they avoided breaking laws on international transport of endangered species. They found that all but three of their purchases had almost certainly been illegally hunted, imported, or processed.

The problem of fraudulent marketing of seafood is not limited to whales. Red snapper is popular but getting harder to find because of overfishing. A college class with their professor analyzed fish from vendors around the country and found that deceptive advertising was rampant, with 75 percent of the fish incorrectly identified. High school students showed that some tuna sold in Manhattan was really tilapia. Mislabeling allows dishonest seafood purveyors to get top dollar for lower quality fish, and worse, gives the false impression to consumers and resource managers that desirable fish remain abundant.

TV crime shows like *CSI: Miami* have brought genetic testing into the households of millions. These same techniques help scientists identify known species and discover new ones, and consumers are already benefiting.

◀ A close-up of the eye of a sperm whale *(Physeter macrocephalus)*. These whales were prized for a waxy material used in candles, soaps, and cosmetics.

FAST FACT:
More than 50 countries are barcoding their biodiversity. See *www.barcoding .si.edu*. In the plankton, where small floating animals can be hard to tell apart, more than 2,000 species have barcodes. Barcodes can be used to count numbers of species, even those without names. Getting a barcode should soon not be much harder than taking a photograph.

FEATURED: *Red Snappers, Whales*

◀ Two-spot red snappers *(Lutjanus bohar)* with a school of slender fusiliers near Fiji. When large, these snappers may cause ciguatera poisoning if eaten.

▲ This minke whale (*Balaenoptera acutorostrata)* is probably safe from whalers while swimming in waters off the coast of Washington.

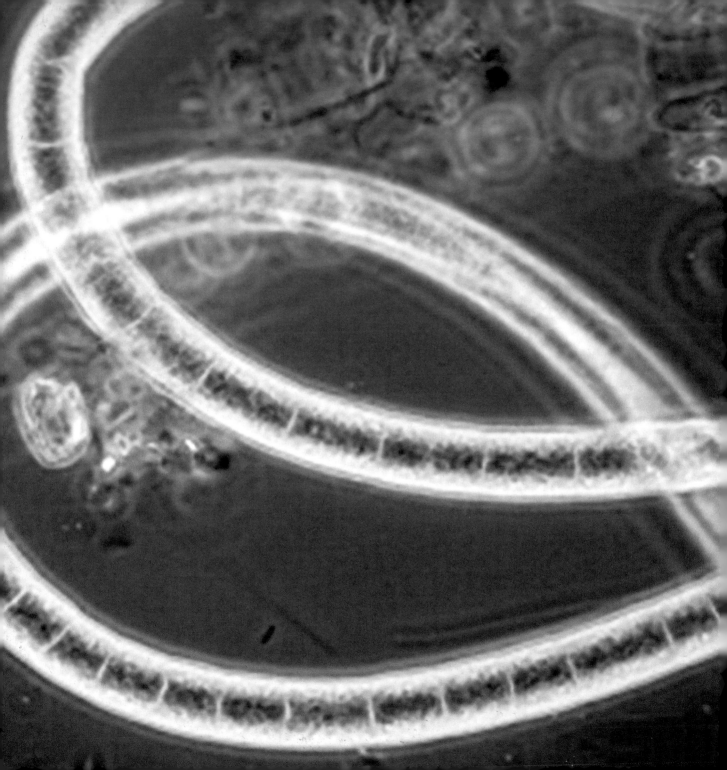

FEATURED: *Blue Whales, Mega-Bacteria*

THE SEA IS BIG AND SO ARE SOME OF ITS CREATURES. The blue whale is the largest animal ever to live on the planet, with a maximum length of about 100 feet and a maximum weight of about 200 tons. Ocean animals have the advantage that water supports their weight, so it is no surprise that the blue whale is more than twice the weight of the largest dinosaurs. This also explains why giant kelps can grow 100 feet tall without any wood to support them.

Although bigger fish often do eat little fish, and sea monsters swam through the nightmares of sailors for centuries, the very biggest sea animals are not predators. The blue whale filters plankton, and so do the two biggest sharks—the whale shark and the basking shark. On land, the same patterns hold—elephants and giraffes and giant tortoises are all vegetarians, as were the giants among the dinosaurs. For the largest animals there probably just isn't enough food to support living as predators at the top of the food chain.

Even the biggest microbes are found in the ocean. Mega-bacteria have been found growing in the guts of unicornfish, amid the shipwrecks of the Namibian coast, and in oxygen-poor mud off the coast of Chile. Getting big is hard for a single cell, and some of these giants accomplish the feat by copying their genes thousands of times—the bacteria in the unicornfish have more than 40 times the DNA of the typical human cell. Giant viruses, or giruses, bigger than the smallest bacteria, are also common in the ocean.

But bigger is not always better, and so the ocean is also home to some of the smallest known creatures. The world's smallest fish (or close to it) lives on coral reefs—it is less than one-third of an inch long and weighs in at a mere 7/100,000 of an ounce. On beaches, hundreds of species are small enough to crawl easily between grains of sand.

If you're looking for new entries, big or small, for *Guinness World Records,* the ocean is clearly the place to go.

▲ A blue whale (*Balaenoptera musculus*) dives off the western coast of Costa Rica. These whales are found in all the oceans of the world.

FAST FACT:
The deep sea has some surprisingly large species, especially considering that food is often scarce. In addition to the famous giant squid, sea spiders can be the size of dinner plates, and shrimplike creatures (related to the backyard roly-polies, or pill bugs) can reach more than a foot in length.

HOW MANY?

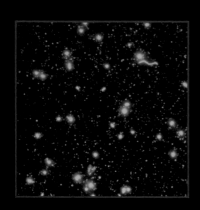

▲ The microbes in a drop of sea-water look like stars in the night sky. The small dots are viruses, and large ones are bacteria.

FAST FACT:
Forty years ago, no one knew there were so many bacteria in the ocean. The vast majority refuse to grow in the laboratory, so scientists assumed that the clear blue water they collected was largely lifeless. Even the microbes that produce much of the oxygen we breathe were discovered only recently, in the late 1980s.

THE QUESTION, "HOW MANY?" LIES AT THE HEART OF any census. Yet it is a deceptively simple question because it is often impossible to count everything. This is particularly true for the tiny but numerous microbes. Instead, we census a small sample of seawater (a relatively easy task), and do some arithmetic to get an estimate for the entire ocean.

The amount of seawater in the ocean is huge—more than 300 million trillion gallons. And the number of bacteria and other similar-size cells (which we'll lump with bacteria for simplicity) in just a drop of seawater is also enormous—between 2,500 and 350,000. Multiply numbers by volume and you get the staggering figure of more than 100,000,000,000,000,000,000,000,000,000,000 ocean bacteria. (The number of stars in the universe is dwarfed by comparison—you would need to take away eight zeros.) And for every one of these bacteria, there are about ten viruses, and for every 50 bacteria there is another larger single-celled organism as well. So, although we tend to ignore what we can't see, ignoring the ocean's microbes is not an option—they dominate the ocean, and consequently the Earth. Every other molecule of oxygen that we breathe comes from one of them.

Censusing animals can be much harder, especially big ones, because you can't just count what's in a cup of seawater and extrapolate. Even for whales, which come to the surface to breathe, censuses are a challenge. Survey ships traverse the seas, and the ratio of new sightings to resightings of known whales (identified by their markings) helps estimate their numbers. But sightings are few and far between—in Antarctica, from 1988–2001 only 120 blue whales were spotted in more than 170,000 miles. Of course there are many whales unsighted and many more fish than that. But even in weight, animals add up to surprisingly little when compared with the mighty microbes. For every pound of animal flesh, ranging from the tiny shrimplike members of the plank-

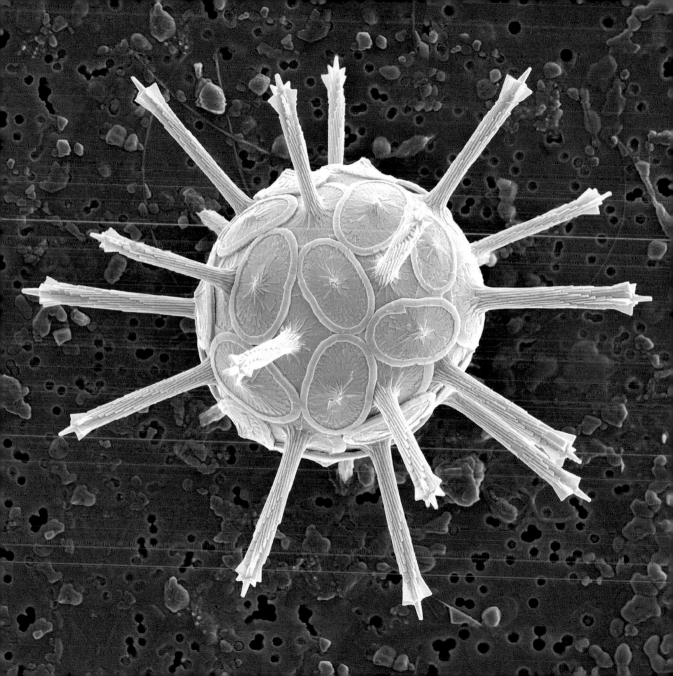

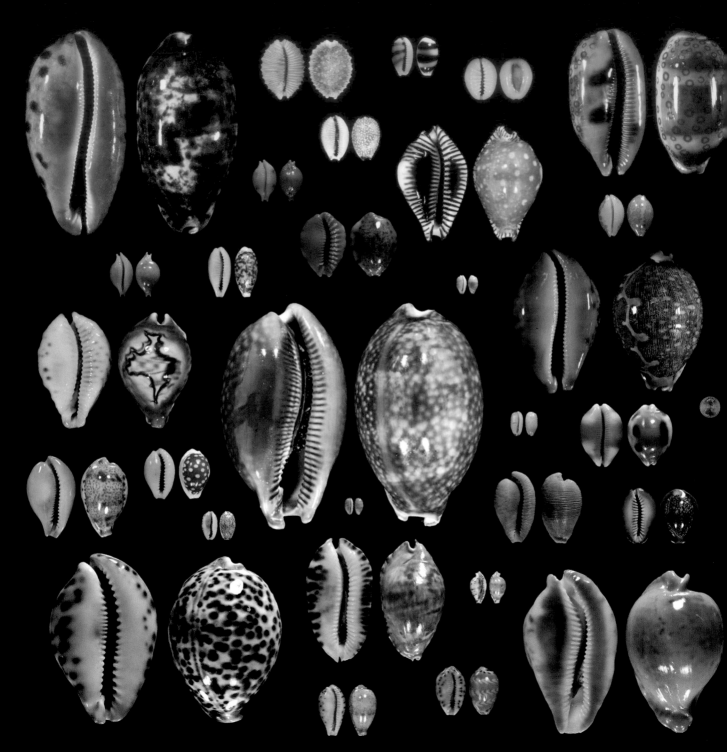

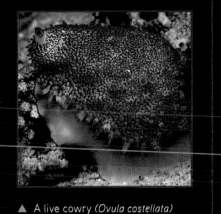

FEATURED: *Microbes, Cowries*

WHEN ASKED WHAT GOD WAS LIKE, AN EMINENT biologist of the 20th century is said to have answered that, based on his creations, he must have been fond of beetles. With today's knowledge, however, a biologist would probably answer that God was a fan of microbes.

Microbes are so diverse that we can't begin to count them all. Viruses mutate like mad, making them the most diverse, although some argue that these snippets of genes that need other organisms to come to life don't really count as proper organisms. But the other microbes, which do count, are also wildly diverse. There may be one billion types of bacteria in the ocean, and 20,000 in just a quart of seawater.

The tree of life has three big trunks—Bacteria, Archaea (once lumped with Bacteria, but in some ways closer to us), and Eukarya (which have cells like ours). But even most Eukarya also have just a single cell. The familiar large organisms—animals, plants, and fungi—represent just three tiny twigs on life's tree.

Life began in the sea, and most of the 30-odd major kinds of animals are still found there. Some are familiar (even if their names are not)—the mollusks (like clams, snails, and squids), arthropods (insects on land and shrimps, lobsters, and crabs in the ocean), worms, sponges, cnidarians (such as corals and jellyfish), and echinoderms (which include sea stars and sea cucumbers). But many marine groups are unfamiliar, sometimes even to most marine biologists.

The land does have more species than the sea, but animal life in the ocean comes in so many forms that we have no idea whether the total number is closer to two million or ten million—or more. Some can be easily told apart only by their DNA, most are rare, and the ocean remains largely unexplored. But coral reefs are clearly the rain forests of the sea—up to one-third of all ocean species live on reefs, even though if reefs were squashed together they would all fit into the area of Texas or France.

▲ A live cowry *(Ovula costellata)* pulls its colorful "skin" over the top of its shell, keeping it smooth and clean.

FAST FACT:
At the opposite extreme, the least diverse major group of ocean organisms is the Placozoa, with just one species (though more are likely to be found using genetics). These simple smudges of tissue were first discovered in an aquarium. There are also only a few species of cycliophorans, tiny creatures to date found only on lobsters.

◄ With more than 200 species, cowries are one of the most diverse and beautiful groups of snails. The shells of some were once used as money.

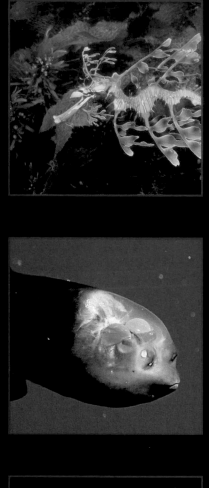

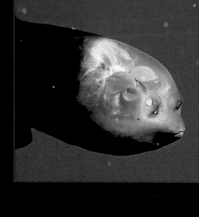
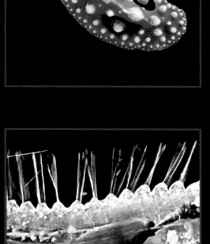
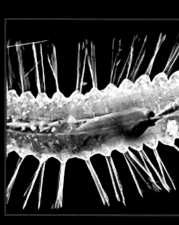
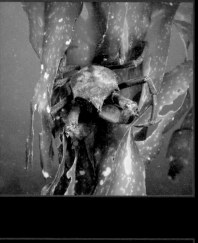
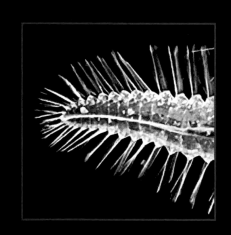

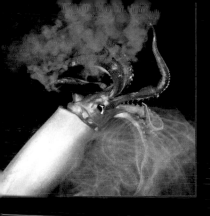

2

APPEARANCES ARE EVERYTHING

Marine organisms have many strategies for getting noticed or keeping a low profile. Markings that help an animal blend in can keep it safe from predators, but bright colors can also protect by sending out a clear warning: "Don't mess with me." Trickery is always an option—harmless animals may masquerade as dangerous, and decoys may allow a quick escape. Sometimes, regardless of reason, ocean life looks just plain weird.

BLENDING IN

PREDATORS BLEND IN TO SNEAK UP ON THEIR PREY, AND prey blend in to avoid being spotted. The stakes for the prey are usually higher—with failure, the predator loses its dinner, but the prey loses its life. But small fish are often both predator and prey, providing a doubly powerful advantage to being inconspicuous.

On land, insects mimic sticks and leaves with extraordinary precision, and under water, equally striking animal mimics of sea grasses and seaweeds can be found. Sometimes, the mimics not only look like plants but also act like them, waving back and forth with the currents. Other sources of evolutionary inspiration are sponges, corals, sea fans, rocks, and sand. Anything that looks unthreatening to a potential prey or inedible to a potential predator is fair game in the contest of camouflage. The success of mimicry can even be a real hazard to humans—more than one diver trying to steady him- or herself has been impaled on the deadly spines of a stonefish, thinking that it was, as the name suggests, a stone.

Crabs have evolved a spectacular array of camouflage strategies to protect their appetizing flesh from would-be consumers. Some look just like the seaweeds to which they cling. Crabs also have one advantage that fish don't have—they can hold on to objects, such as sponges, with their legs or attach seaweed and other debris to hooks on their shells.

Like beauty, camouflage is in the eye of the beholder. Red appears brown or black in all but the shallowest water. Penguins stand out when they're on land, but when they're swimming, their white bellies and black backs hide them from predators gazing up toward the sky or down toward the depths.

Such exquisite details of animal camouflage were once used as an argument against evolution—the matches appeared too perfect. But appearances are everything when it comes to natural selection and the struggle to eat and not be eaten.

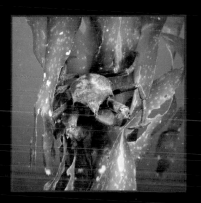

▲ This kelp crab (*Pugettia producta*) eats kelp, lives in kelp, looks like kelp, and even has a few hooks for attaching kelp to its body.

▼ Decorator crabs sometimes go to great lengths to make themselves inconspicuous. Seaweed is the most common cloak, but some crabs wear a sponge hat.

FAST FACT:
Coral reef fish aren't as conspicuous in nature as they are in an aquarium. From a distance they blend in surprisingly well against the bright colors of a healthy reef. But reef fish stand out against seaweed-covered dead coral, so they are more likely to get eaten if the reef is in bad shape.

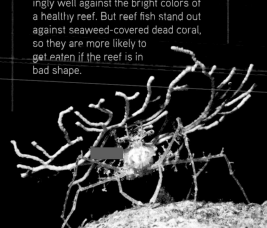

◄ Thanks to its leafy extensions, this sea dragon (*Phycodurus eques*) hides in plain sight among seaweeds off the coast of southern Australia.

▲ Though these Antarctic gentoo penguins (*Pygoscelis papua*) may not seem camouflaged on land, once they get in the water their two-tone tuxes hide them from predators.

MASTERS OF DISGUISE

▲ A close look at the eye of this giant Pacific octopus *(Enteroctopus dofleini)* shows the tiny pigment spots that together create the octopus's pattern.

FAST FACT:
In the competition between flounders and octopuses for master of disguise, the octopuses clearly win, both in speed and accuracy. But a few octopus species take things one step further by mimicking the flounders themselves, not only their shape and color but even their distinctive swimming style. Mimic mimics mimic!

B LENDING IN AGAINST A SINGLE BACKGROUND TO become invisible is hard enough, but some animals can do far more, switching disguises depending on where they are. On land, chameleons are renowned for their ability to change color, but they are amateurs compared with squids, octopuses, and cuttlefish, which can change their "costumes" to perfectly match their new backgrounds in a matter of seconds. How do they do this so well and so quickly?

The secret is great skin. Across the surface of an octopus is an array of small sacs of colored pigments, each independently controlled by tiny muscles with a direct nervous-system connection to the brain. Open the sac and the area of pigment expands, making the skin darker. Close the sac and the color disappears. Even a small octopus, with a body the size of a human fist, will have thousands of color sacs that can be opened or closed in one-hundredth of a second. And the skin's surface itself can be smoothed out or puckered up, like goose bumps on steroids, to match the texture of the surroundings.

But how does the octopus know which disguise to put on? Scientists have explored this question by giving animals different kinds of checkerboards to match. Depending on the size of the squares, the octopus turns on its uniform pattern (against very tiny squares), its mottled pattern (against medium-size squares), or its disruptive pattern (against large squares). Uniform and mottled patterns mimic the background, while disruptive patterns make it hard to tell where the octopus ends and the background begins. Just these few simple matching rules appear to underlie the octopus' remarkable abilities.

Octopuses and their relatives are not alone in their ability to disappear into the background—the flounder also has a sophisticated vanishing act, though in fish, pigment cells are controlled via chemicals released into the blood rather than by nerves directly, and the transformation usually takes longer. But regardless of method, these underwater disguises put most Halloween costumes to shame.

▶ Now you see me! Now you don't! This flounder (*Bothus* sp., above) and octopus (*Octopus vulgaris*, below) blend into the underwater scenery.

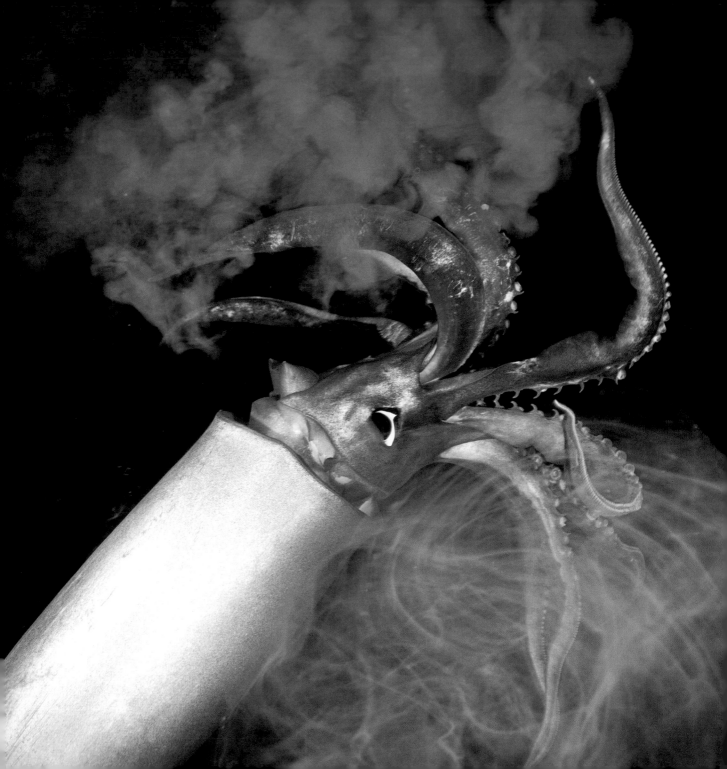

DECEPTION & DISTRACTION

HESITATION CAN SPELL THE DIFFERENCE BETWEEN A good meal and an empty stomach, so markings like fake eyes may help the vulnerable deceive or distract potential predators, even if only briefly. Those of the aptly named eyespot skate seem to make it look large and dangerous. Those of the four-eyed butterflyfish, which has an extra set of "eyes" near the tail, could provide confusing signals about likely escape directions. Eyespots on fins may direct a predator toward more disposable body parts, or if flashed when fins open, they may startle a predator and buy a few critical extra seconds. Fish are not the only animals with eyespots—swimming crabs show off large fake eyes on their back legs when cornered or chased, and some octopuses also flash eyespots when threatened.

A truly successful distraction leaves a predator not with a piece of flesh, but with nothing at all. Squid ink is designed to do just that. If camouflage fails, a puff of ink hanging in the water can confuse the enemy for just long enough to make a quick escape possible. Some inks are released as clouds and function as smoke screens. Others are held together with mucus, creating a phantom squid to distract the predator. One particularly clever squid maneuver involves changing color and releasing ink at the same time, followed by a quick exit. But ink is black and so it's not much use in the dark of night or in the deep sea where many squids live. There, glow-in-the-dark ink provides the solution for at least one of them.

Like many good ideas, Mother Nature (aka evolution) has thought of it more than once. The recently discovered worm named *Swima bombiviridis* (roughly translated as the "swimming green bomber") has reinvented the squid's trick. This deep-sea dweller wears a necklace of liquid-filled sacs. If threatened, several are released and instantly glow bright green, allowing the worm to slip safely away from the puzzled predator. As any good magician knows, distraction underlies every successful sleight of hand (or in this case neck).

▲ Borrowing from the squid's repertoire, the green bomber worm *(Swima bombiviridis)* releases sacs of glowing green fluid from its neck if threatened.

◀ Just like a magician disappearing in a puff of smoke, this squid *(Dosidicus gigas)* distracts its predator with a cloud of ink before making its exit.

STANDING OUT

▲ On a reef near Australia, the colors of this sea slug *(Chromodoris elizabethina)* serve as a billboard that reads, "don't feed on me!"

FAST FACT:
Although many ocean animals produce large numbers of tiny, tasty young that spend a long time floating in the plankton, others produce fewer, larger young. Though these tend to settle to the bottom quickly, they are still at risk of being eaten. Often, however, they're brightly colored, and as expected, taste terrible to fish.

THE ADVANTAGE OF BLENDING IN IS OBVIOUS—IT allows an organism to escape the attention of its predators. But the opposite approach also works. Wasps use their bright black and yellow stripes to advertise "don't mess with me," and poisonous or distasteful animals, like poison arrow frogs or ladybugs, are also often hard to miss.

The same holds true in the sea. Slugs lack a shell and move slowly, making them potentially easy prey. For protection, some hide by perfectly matching what they eat, but others rival the brightest coral reef fish with their blues, reds, greens, yellows, and blacks. Many of these slugs feed on sponges and acquire the nasty chemicals that the sponges use to fend off predators. They are, literally, what they eat, and one unpleasant mouthful is usually enough to make a long-lasting impression on any fish that tries to eat a sponge-eating slug.

Why advertise when one can blend in? Warning colors work because the predator either instinctively avoids or quickly learns to steer clear of brightly colored prey. The learning route is a bit tricky at the outset because it requires that the prey survive being sampled. Fortunately, many slugs do seem to survive being eaten and then spit out. The worse they taste and the more conspicuous they are, the more likely is the predator to remember the lesson and avoid them in the future. In some places, lots of different slugs have converged on the same color pattern, making it easier for predators to learn to avoid them. This opens the door for perfectly delicious mimics to take advantage of the scheme—some small, tasty fish do very good slug impersonations.

But what of delicate prey that have little chance of surviving a misguided predator attack? It's possible that being really conspicuous might actually make an animal less vulnerable. Just as blue rice might not be a popular buffet item, fish seem to be similarly wary when it comes to weird-looking food.

▶ Coral reefs are full of hungry creatures, but the coloration of this poisonous sea slug *(Phyllidia ocellata)* warns predators that it is off the menu.

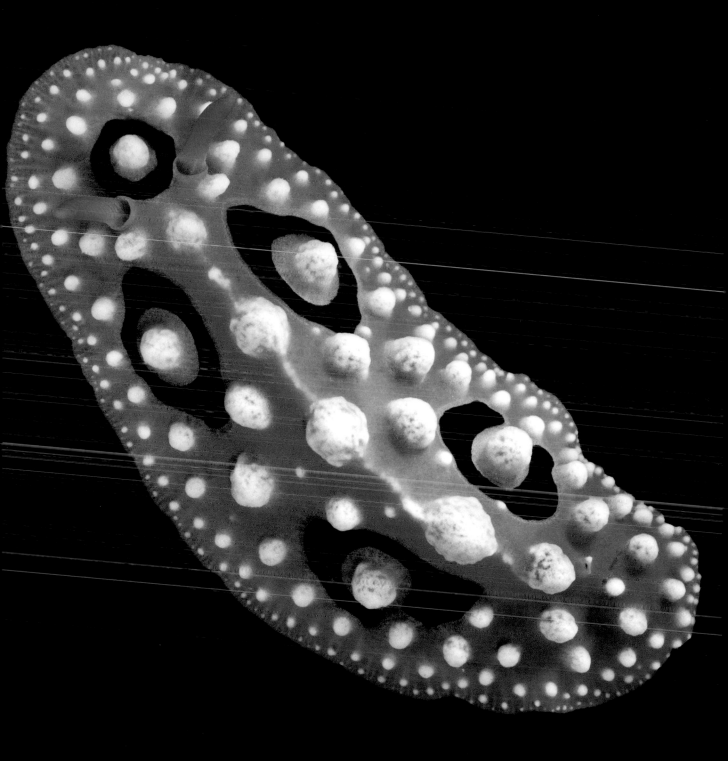

STRANGER THAN FICTION

FEATURED: *Barreleye Fish, Slipper Lobsters*

THE DEEP SEA CONTAINS MUCH THAT IS BEAUTIFUL AND much that looks bizarre, at least from our perspective, living as we do in a dry, light-filled, and low-pressure world. Among the creatures that seem stranger than fiction is the barreleye fish.

Barreleyes get their name from their long tubular eyes. In low light, a large aperture makes it easier to capture an image, just as it does with a camera. But a large aperture on a spherical eye results in an eye so big that it takes up too much of the head, so tubular eyes are one solution for keeping the aperture large but the eye volume small. Barreleyes live 2,000–2,600 feet deep, where the only light comes faintly from above or from the bioluminescence of other deepwater creatures. Eleven other families of dark-dwelling fish have tubular eyes, as do the nocturnal owls and bushbabies on land.

Tubular eyes have one great limitation, and that is their field of view. Much like the telescopes they resemble, the tubes need to be pointed in the right direction. Owls and primates can turn their heads, but barreleyes were once thought to have their eyes permanently pointed upward (their green color filters out sunlight and makes bioluminescent organisms more conspicuous). Upward-pointing eyes may be excellent for scanning for food above, but they pose a challenge for getting food into the tiny mouth, which is out of the field of view (what look like forward-directed eyes are actually nostrils). New videos show that barreleyes can actually rotate their tubular eyes, which solves the problem of mouth-eye coordination.

But new videos reveal something even stranger. Over the movable eyes, scientists saw for the first time that the entire top of the head is transparent. A large clear shield (usually lost when these fish are captured) covers a chamber filled with see-through fluid that surrounds and protects the eyes. Why the fancy eye gear? It may be that these fish steal food from the tentacles of jellyfish, and the shield protects their eyes from getting stung.

▲This baby slipper lobster is completely transparent before growing a thick shell. Its bizarre eyes may confuse predators while it floats in the plankton.

FAST FACT:
Transparency seems weird to us because it is rare on land (glass frogs and insect wings are two examples). But many small, tasty animals floating in the plankton wear what amounts to an invisibility cloak. The stomach and retina are obvious design challenges, but so is the skin, where tiny bumps help (as designers of stealth bombers know).

◄ What looks like the eyes of this barreleye fish (*Macropinna microstoma*) are actually nostrils. Its green tubular eyes are protected by a transparent shield.

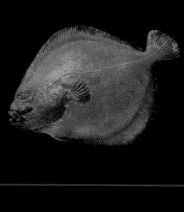

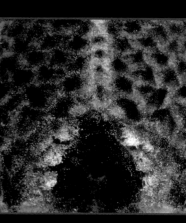

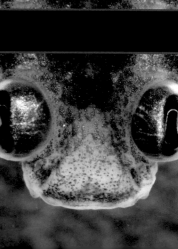

3

SENSE & SENSIBILITIES

For most humans, "seeing is believing," and speech depends on hearing. Smell, taste, and touch also help us keep track of, manage, and enjoy our environment. But the ocean and air behave very differently when it comes to light, sound, and odors. Some ocean animals are also able to detect and use magnetic fields. Making sense of sensory data takes brainpower, and the mental prowess of marine animals can be startling.

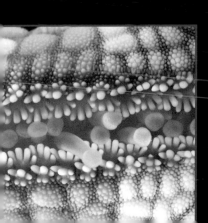
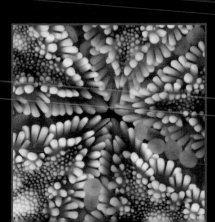

SEEING THE LIGHT

▲ This four-eyed fish *(Anableps anableps)* from South America can see above and below the surface at the same time with its unusual eyes.

FAST FACT:
Two Australian damselfish species can be distinguished only by the markings on their faces, which reflect ultraviolet light. Most fish predators can't see ultraviolet light, so these markings allow the damsels to communicate covertly. Each male has a harem, and subtle differences in the markings probably help him tell his damsels apart.

SUNLIGHT CHANGES DRAMATICALLY AS IT PASSES through seawater. Deeper seascapes appear bluer than shallower ones because the long wavelengths of red light are filtered out first. Eventually, sunlight is completely blocked, but even in very deep water there is light, thanks to the bioluminescence that the animals themselves produce.

The stalked compound eyes of the predatory mantis shrimp constantly scan their surroundings, seeing many things that we cannot. Humans have three types of color detectors that are tuned to different parts of the rainbow of wavelengths, from red to blue, that we see. But many mantis shrimps have ten within that range, plus another five or six that can see ultraviolet, to which we are blind. They can also recognize polarization patterns, in addition to color patterns.

Most animals cannot see as well as mantis shrimps, but they have nevertheless invented some ingenious optics. Some shore-dwelling fish need to see clearly both in and out of water. We need a mask to do this, but the four-eyed fish succeeds by splitting each cornea, pupil, lens, and retina into two parts. It swims with its eyes half submerged— the top part is designed to see in air, the bottom part to see in water. Deep water also has its four-eyed animals. Spookfish have two large eyes that look upward to see silhouettes and two small eyes that look downward to see flashing bioluminescence.

Eyes don't always need much brainpower to back them up. Some clams have hundreds of eyes that function as burglar alarms, causing the shells to snap shut if a shadow of danger is spotted. Swimming scallops have the best vision of all clams, with an array of eyes on tentacles that can both find shelter and detect enemies. But even the utterly eyeless sea urchins can see. Light sensitivity is spread across their surface, and they use their spines to break up the light into meaningful packages. Though not in the mantis shrimp's league, the body of the sea urchin functions as one giant compound eye.

▶ Compared with the array of colors seen by the compound eyes of this mantis shrimp *(Odontodactylus scyllarus),* human vision is practically in black and white.

MUSIC TO THEIR EARS

FEATURED: *Drum Fish, Grunt Fish, Whales, Dolphins*

SEAWATER IS MUCH FRIENDLIER TO SOUND THAN TO sunlight. Light gets filtered out by water and scattered by floating particles, so being able to see distances of even a few hundred feet is unusual. In contrast, sound travels more than four times as fast and much farther in the ocean than it does in air, which is why we use sound to map the seafloor rather than light.

Drums drum, croakers croak, and grunts grunt. Many fish, and a few other creatures, such as snapping shrimp, make sounds. Sounds are often used to intimidate rivals or attract mates—after all, it takes a larger body to produce lower-pitched and louder sounds. The most complex social sounds are made by marine mammals—some, like the song of the humpback whale, are eerily appealing to us as well. (The Voyager spacecraft even carried their recordings into outer space.) But the most extraordinary ocean acoustic performance is echolocation—here the sound producer and the intended listener are one and the same. Toothed whales and dolphins started using echolocation—also known as biosonar—to find prey about 35 million years ago.

Beaked whales dive deep to feed on squids and fish. While they are looking for food, they emit clusters of clicks, one every four-tenths of a second. When they are about a body length away from dinner, the click cluster is replaced by a softer buzz, two to five seconds long, composed of clicks emitted as frequently as every five-thousandths of a second. This seems fast, but the whales always wait until they hear the echo from one click before making another, in fact a bit longer—probably to make sense of what they have heard or because they can't click any faster. A whale emits about 15,000 sonar clicks during one dive, and prey can probably be detected at a distance of 900 feet. Many targets are ultimately ignored, suggesting that the detailed "sound pictures" allow them to be choosy. Although one bat also uses echolocation to find fish at the surface, the toothed whales and dolphins are the real bats of the sea.

▲ This highhat (*Equetus acuminatus*) is a member of the drum fish family whose sounds are especially loud during mating season.

FAST FACT:
Swim bladders are gas-filled bags that fish use to keep from sinking or floating, but some fish also vibrate them to produce a song. Though the songs are hardly of nightingale quality, fish, like birds, use their songs to repel intruders or attract mates. Underwater listening devices can also let us know what the fish are up to.

◀ Grunt fish (*Haemulon* sp.) are named for the sounds they make both in and out of the water by grinding their teeth together.

THE NOSE KNOWS

▲ There are hundreds of suckers on the underside of this sea star (*Nardoa rosea*), which allow it to taste, smell, and hold on.

FAST FACT:
As we change ocean chemistry by burning fossil fuels, we also affect the ability of animals to navigate by their noses. Baby anemonefish, when put in acidic seawater like that predicted for the end of the century, can no longer recognize predator and prey or good places to settle—bad news for future Nemos.

A GOOD SENSE OF SMELL CAN BE INVALUABLE IN FIND-ing food, a mate, or a home. Albatross can smell fish from long distances, polar bears can track seals by their odors, and baby reef fish can recognize their own reef's water. Exactly how they do this remains a mystery, but they are certainly well equipped for the task. Fish have tens to hundreds of distinct genes for sensing different chemicals, and birds and mammals have even more.

Pacific salmon are the quintessential example of exquisite nose-based navigation. The young salmon hatch upstream in rivers but migrate down to the sea to finish growing up. After spending several years in the open ocean, they return to freshwater to reproduce and start the cycle anew. And they return not just to any freshwater, but to the very stream where they started out. Different streams have different aromas, and a fish memorizes the smell of its own stream on its way to sea and carries the memory for years before heading back. The system isn't perfect, but it's remarkably close—about 95 percent of the fish that make it back to some river will find their birthplace.

Why should salmon go to such lengths to mate where their parents did? Although it might seem silly to pass up perfectly good streams in a search for their ancestral home, fish have no way of knowing if a stream really is perfectly good—except the home stream, which by definition is good, since they survived.

For people, taste buds and their nasal equivalents are quite different in the way they work, but at the level of the brain the messages are mixed, which is why food seems so tasteless when you have a cold. In the sea the distinction between smell and taste is even more blurred, especially for the many creatures that don't have proper noses, or for that matter, tongues. But that doesn't mean they don't pay attention to the molecules around them. The antennae of crustaceans and the suckers of sea stars are covered with cells for tasting and smelling. "Better living through chemistry" is their motto.

▶ These sockeye salmon (*Oncorhynchus nerka*) have migrated from the Pacific Ocean back to Russia's Kamchatka Peninsula to spawn in freshwater.

KEEPING IN TOUCH

I F A CAR PASSES NEARBY WE FEEL IT, BUT WE DON'T USU-
ally feel the movement of small objects. In water, which is much
denser than air, the movements of even little fish can be sensed
as perturbations to the flow. And unlike a car, which we stop feeling as soon as it leaves the scene, underwater trails of altered
flow last much longer and so can be followed.

Many ocean animals can sense flow—lobsters, for example, can detect water movement with hairs on their claws and
tail. Keeping track of water movements is serious business
for fish, and they have an elaborate sensory system devoted to
this task. Some sensors at the surface of the skin detect water flow,
and others are slightly sunken and register changes in pressure. This
row of sunken sensors can be clearly seen on the sides of most fish—
hence its name, the lateral line. Without the lateral line, coordinated
swimming in a school would be impossible. It works so well that engi-
neers have built artificial lateral lines for possible use on submarines.

Marine mammals don't have lateral lines, but they have something
that is almost as good—whiskers. Seals, sea lions, and walruses have
the most sophisticated and extensive set of whiskers of any mammal.
Some seals use their whiskers to follow swimming prey. Because the
wake of a fish resembles a ladder of small whirlpools and lasts for several minutes, the seals can use their whiskers to follow the trail. The
existence of blind but well-fed seals in the wild is a testimony to the
effectiveness of their whiskers.

Bearded seals and walruses hunt stationary food, such as clams, on
the ocean bottom. Their whiskers are even more elaborate, so that in
dark, turbid water they can feel exactly what they are eating. The average bearded seal has 244 whiskers connected by more than 320,000
nerve cells. These seals can tell the difference between objects by
using their whiskers as well as monkeys can by using their hands!

▲ Clearly running from its gills to its
tail, the lateral line, found on all fish, is
especially conspicuous in this rough-
scale sole *(Clidoderma asperrimum)*.

FAST FACT:
Seals are not the only beneficiaries
of whisker wisdom. Differences in
chemical composition along the length
of the whisker tell us where seals
have gone (from Antarctica to the
subtropics) and what they have eaten
(from krill to fish). Even the wavy
profile of individual whiskers suggests
new designs for devices to detect
water flow.

ANIMAL MAGNETISM

▲ In the waters of the British West Indies, a loggerhead sea turtle *(Caretta caretta)* navigates while a suckerfish catches a ride.

FAST FACT:
Many animal navigators use more than one cue, combining sights, sounds, and smells to help them tell direction. By combining a swimming pool with a planetarium, scientists have found that harbor seals may even be able to navigate by reading the stars, much as the Polynesians did as they moved across the Pacific.

BEFORE THE INVENTION OF SATELLITE-BASED GPS, NAVIgation was an art as well as a science. For direction you needed a compass, for latitude a sextant, and for the hardest, longitude, you needed a good timepiece. It wasn't until the 18th century that explorers had all the tools needed for reliable open-ocean navigation, but animals have been doing this for many millions of years by reading Earth's magnetic field.

Loggerhead turtles are champion navigators in both the Atlantic and Pacific. Hatchlings from Florida ride the Gulf Stream to the North Atlantic gyre, the clockwise current that surrounds the Sargasso Sea. They spend several years in the gyre, sometimes following it to near the north coasts of Africa or South America. One might think that such a broad area would require little navigational precision, but accidentally leaving the gyre would be fatal. Near Portugal, for example, a northward current splits off, and any turtle unfortunate enough to wind up in it would be carried to a chilly death. Fortunately, Earth's magnetic field varies in predictable ways, and young turtles can sense the longitudinal and latitudinal patterns. A young turtle taken from the southern part of the gyre tends to swim northwest; one taken from northeast of the gyre heads south.

There are also spikes in the magnetic field caused by concentrations of rocks containing magnetic minerals. Animals capable of learning the positions of these magnetic irregularities could use them as landmarks. Hammerhead sharks in the Gulf of California seem to do this, following magnetic contours at night to and from a seamount.

The list of marine animals that respond to magnetic fields continues to grow—small shrimplike creatures, sea slugs, and even bacteria are on it. Magnetic material has been found in migrating eels, as well as spiny lobsters, which also seem to navigate using a magnetic map. But we still know almost nothing about how animals actually use their

OUR COGNITIVE COUSINS

WHAT MAKES PEOPLE SO SPECIAL? TOOL USE, SELF-consciousness, language, and culture are high on the list, but in fact all of these characteristics can be found elsewhere in the animal kingdom. *Homo sapiens* and other primates are close relatives, so it is perhaps not surprising that chimpanzees use tools or that capuchin monkeys have a sense of fair play, even rejecting cucumbers (which they normally accept) when they see their neighbors getting grapes. But the qualities that we often think of as uniquely human evolved not just on land but in the ocean as well.

Among the invertebrates, octopuses are known for their intelligence, even exhibiting evidence of playfulness, tool use, and personality. But these skilled predators are solitary and consequently lack culture. Dolphins, on the other hand, are large-brained, long-lived, group-living predators, and it is here that we find the greatest parallels with human culture and cognition.

Culture depends on the ability of animals to pass on things they have learned to others. Many animals have culture in this sense, but what sets dolphins apart is *what* they pass on. Some bottlenose dolphins carry sponges on their beaks that they use as tools to ferret out fish from the bottom. This ability is handed down through the generations (especially in females), with some families—grandmother, mother, and daughter—all feeding in this highly specialized way.

Another characteristic that dolphins share with humans is their ability to recognize themselves in a mirror. When confronted with a mirror, most animals behave as though they are interacting with another individual, and even in humans, the ability to recognize that the image in a mirror is oneself does not occur before the age of 18 months. Dolphins not only recognize themselves, but if a black mark is put on the body of a dolphin, it will spend extra time at the mirror to look at the mark.

◀ Bottlenose dolphins *(Tursiops truncatus)* are highly intelligent and teach valuable skills to other members of their group.

▼ A newly hatched baby octopus escapes from its ruptured egg case, leaving its siblings behind. As an adult, it will be a clever predator.

FAST FACT:
Female sperm whales live in groups. Mothers must dive deep and long to find food, and they share diving and babysitting duties so that each calf has an escort until Mom returns. This could be an example of "I'll scratch your back if you'll scratch mine," a hallmark of human societies.

FEATURED: *Octopuses, Dolphins, Sperm Whales*

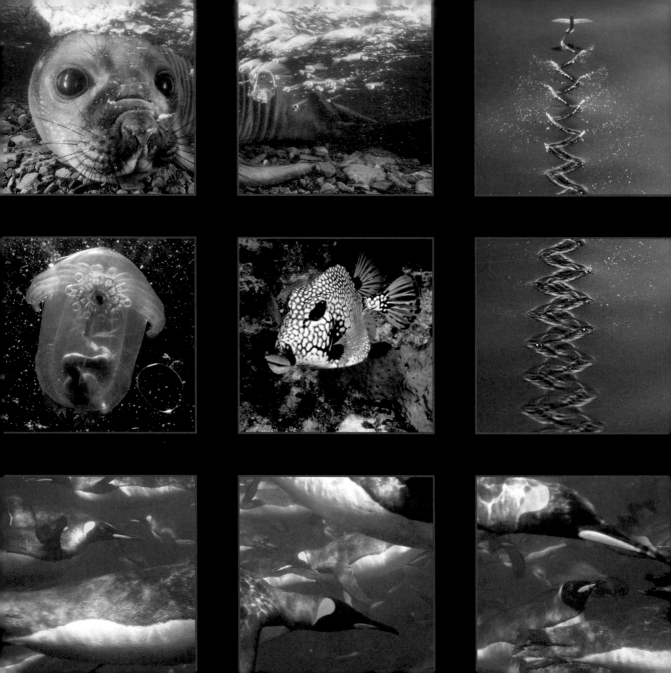

OCEAN LOCOMOTION

The ocean is vast, and many ocean animals need to travel at some point in their lives. Most fish and some birds swim, while many birds and a few fish fly and walk—the most mobile may cover thousands of miles each year. Some, both large and small, move up and down to find food or avoid being eaten. Others simply float or attach to floating objects, allowing the currents to direct them. A few even roll from place to place.

FREQUENT FLYERS

WILDEBEEST MIGRATIONS ACROSS THE SERENGETI have captured the imagination of countless wildlife photographers, but the distances traveled are usually less than 125 miles. Oceanic voyagers travel far more and more often—albatross can soar more than 500 miles a day. In terms of effort, the champions are much smaller seabirds that depend on the immensely productive polar seas, feeding in both the Arctic and the Antarctic every year. Tiny tags strapped to their legs that record light, pressure, and depth reveal how remarkable these voyages are.

Sooty shearwaters, or muttonbirds, weigh less than two pounds and feed on fish, squid, and krill. In the South Pacific, during their December to April summer breeding season, pairs nest in burrows. Just one egg is laid, and once it hatches, both parents bring food. They sometimes fish close to home, but they often travel as far as the waters of Antarctica to feed their chick. A 1,200-mile trip to the grocery store and back may seem like a long way to go, but that is just the beginning.

With the coming of fall, the birds head north, traveling as much as 565 miles a day toward one of three winter getaway destinations—western North America, the Bering Sea, or Japan and Siberia. After sharing a burrow for months, mates travel on their own, often overwintering on opposite sides of the North Pacific. Six months later, in October, they head back to their breeding grounds. The trip north is rather loosely timed, with birds arriving over a nearly monthlong period, but the trip south is highly synchronized, with most birds funneling through a narrow corridor over the Equator in ten days.

Millions of sooty shearwaters make this trip every year. With an average life span of 32 years, this adds up to more than one million miles and 6,000 days "on the road." And recent studies of the smaller arctic terns show that they travel even farther over their lifetimes—equivalent to a remarkable three round-trips to the moon! All in a

▲ Each year, the arctic tern (*Sterna paradisaea*) flies an average of 44,000 miles, from the Arctic to the Antarctic and back, the longest known annual migration.

FAST FACT:
Tagging has gone high tech. Ultra-lightweight tags are used on small birds, while tags put inside fish tell their stories after the fish are caught. Other tags send data to satellites when the tag pops off or when the animal surfaces. Some tags simply emit signals—notifying underwater sensors that, for example, "John Doe" has passed by.

FEATURED: *Sharks, Trunkfish, Flying Fish, Mudskippers*

▲ Fish out of water! In the Indian Ocean, a flying fish *(Cypselurus suttoni)* creates a zigzag pattern as it builds momentum to become airborne.

FAST FACT:
Great white sharks migrate great distances, and over the long haul they're probably the ocean's fastest swimmers. California great whites swim out to Hawaii each winter and then return to California in the summer. But on the way back, they take a mysterious break at a location dubbed the white shark café by Census scientists, perhaps for feeding or mating or both.

YOU CAN TELL A LOT ABOUT HOW A FISH SWIMS FROM its shape. Not all fish look like bullets with tails—their bodies range in shape from sticks to balloons. Just as blimps and jet fighters are designed for different kinds of flight, so are the bodies of fish designed for different kinds of swimming. The ability to swim fast for long periods, to accelerate quickly, or to maneuver skillfully all require different designs, and the shape of a fish reflects compromises depending on its lifestyle.

Fish designed for cruising at high speeds have streamlined bodies with crescent-shaped tails that provide most of the propulsion—all other things being equal, the higher the ratio of the span of the tail to its area, the faster the fish can go. Sharks and tunas are utterly unrelated but have converged on this same basic body plan. Large fish with this shape can cross the Pacific or Atlantic with ease.

Other fish mainly use their fins to swim—the single fins on the top and bottom (dorsal and anal fins), the paired side fins (pectoral and pelvic fins), or all of the above. Fins may be rowed like oars or used in an undulating manner. Fin swimming tends to be slower but allows for more agility. At the extreme, trunkfish are basically finned boxes, no good for speed but able to turn on a dime and perfect for maneuvering in the tiny crevices of a coral reef.

Flying fish are built for speed, but they don't limit themselves to the water. They build enough momentum to escape into the air, where they can soar for longer than 40 seconds using their pectoral "wings." This allows them to escape from their predators at a speed much faster than they could ever swim, since water is 800 times denser than air.

A few fish use their fins for walking, both on the seafloor and on land—mudskippers can even climb. Several hundred million years ago, some lobe-finned fish left the water altogether. Today their descendants include frogs, lizards—and people.

▶ A trunkfish *(Lactophrys triqueter,* above) deftly swims among the channels of a Caribbean coral reef, while a mudskipper (below) escapes the water altogether.

TAKING THE PLUNGE

I N EARLY 2010, THE HUMAN DEPTH RECORD FOR FREE-
diving without fins or weights was held by a New Zealander who
reached and returned from deeper than 280 feet in 3½ minutes.
The time record for "simple" breath holding is currently about 11½
minutes. As remarkable as these achievements are, humans are amateurs compared with other air-breathing divers.

Diving birds and mammals use a number of tricks. They store oxygen more efficiently, tolerate oxygen starvation better, and can slow
down their metabolisms—heartbeats as low as four beats per minute
have been recorded in diving seals. In general, the larger the animal,
the slower its metabolism and the more oxygen it can store, so the
deeper and longer it can dive. The slow metabolism of cold-blooded
reptiles like turtles also makes them excellent divers.

Some birds are superb divers despite their small size and warm
blood. King penguins can reach depths of nearly 1,000 feet and stay
under water for seven minutes. Others, like gannets, may not dive for
as long or dive as deep, but they plunge like arrows from heights of
nearly 100 feet and can hit the water at more than 90 miles an hour.

Among the mammals, elephant seals are extraordinary divers—
they have been called surfacers rather than divers because they spend
90 percent of their time submerged. But the champion divers in terms
of depth and duration are the beaked whales, who regularly dive more
than a mile down and often stay deep for an hour.

Human divers, both with scuba and without, can get decompres-
sion sickness, or "the bends." Air is 80 percent nitrogen, and it dis-
solves into the blood and tissues under pressure when diving. Because
nitrogen is not consumed by cells, during the ascent it is released. If
the ascent is too fast, bubbles form, causing sometimes-fatal damage.
Diving animals somehow know their limits, normally, but use of sonar
by the Navy may be scaring beaked whales into making extra dives,

▲ A 400-pound elephant seal pup
(Mirounga leonina) displays its sub-
merging and surfacing skills while it
plays in shallow water.

FAST FACT:
Census scientists have found that
animal divers make great oceanogra-
phers. Especially at the Poles, where
sea ice blocks the view of satellites
and expeditions are costly, recording devices attached to animals can
provide windows to an otherwise
inaccessible world. One southern
elephant seal in Antarctica dove 200
times over a period of 50 days, providing critical data for monitoring our
changing climate.

▲ A flock of gannets (*Morus cap-
ensis*) plunges for fish, entering the
water at near breakneck speeds off
the coast of South Africa.

UP AND DOWN

▼ This arctic copepod *(Paraeuchaeta barbata)* carries its clutch of eggs as it migrates up and down each day.

FAST FACT:
Gulping down plankton is harder than it seems. For the blue whales and their relatives, first the pleated, accordion-like mouth gapes open. Then the tongue inverts into the belly like the finger of a glove, making even more space, and water the volume of a school bus and heavier than the whale itself enters.

WHAT IS THE LARGEST MIGRATION ON THE PLANET? If you were to decide by weighing all the animals involved, then the answer would be the plankton and the animals that eat them. Every day at dusk, throughout the world's oceans, trillions of small copepods, shrimps, jellies, and fish move to the shallows to feed, and at dawn they move back to deeper water. (Some stay put and a few others buck the trend and migrate in reverse.) Why so many move up is no surprise. The small photosynthesizing algae that form the base of the ocean's food chain live where the light is. The slightly larger animals that eat them have to go where their food is, and their predators in turn follow them. But why do the tiny alga eaters abandon their food every morning? During the day, they risk being eaten by their predators, mainly fish, which hunt them by sight. So when the sun rises, they retreat to the safety of the darkness of the deep, since it is better to be hungry than to be some other animal's breakfast. In total, the numbers are so vast that this huge migrating herd of plankton can be "seen" using sonar as a sound-scattering layer that moves up and down.

The total distance that is moved each day may not seem very large—often less than a few hundred feet, depending on the location. But considering the sizes of many of the animals that make this daily trek—often much smaller than an inch—the distance is impressive, as it represents many thousands of times the length of the body. For very small animals, the physics of swimming are also completely different—for them, water is more like molasses, making migration even harder.

Although plankton are by definition small, some of the things that eat them are not. Basking sharks can weigh more than 8,000 pounds, and feed on plankton by swimming with their mouths open. Unless the plankton are dense they will be hungry, so not surprisingly, they too

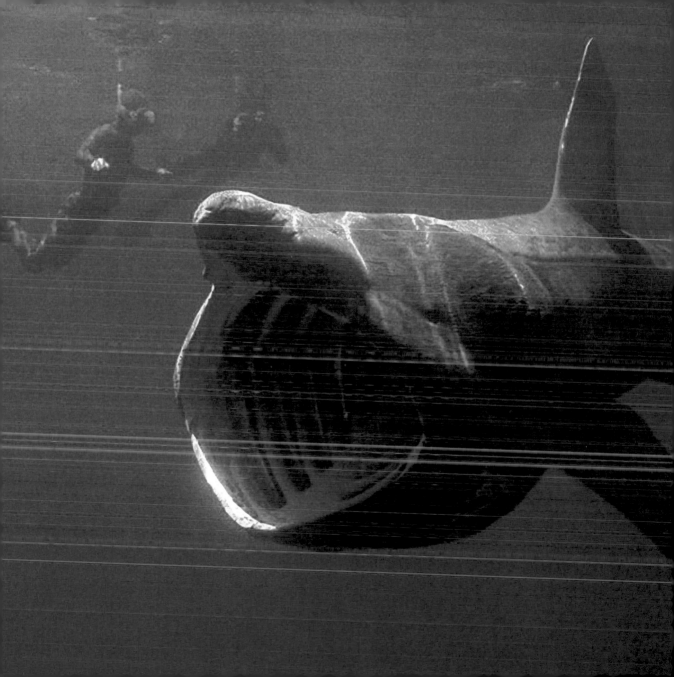

HITCHHIKER'S GUIDE TO THE OCEAN

S OME FISH AND BIRDS AND TURTLES CAN CROSS THE world's oceans unaided as adults, but this requires the ability and energy to move through or above the water. Other animals are fairly sedentary as adults or can't move at all, and they do their traveling as juveniles—tiny larvae that drift, sometimes for thousands of miles, feeding as they go. But not all larvae can feed—so are these marine creatures doomed to spend their entire lives where their parents grew up? Not necessarily, if they can catch a ride on a raft.

Columbus was well aware of rafting—he reported on barnacles growing on floating sticks, and used the presence of foreign objects on the shores of Portugal to make his case for the existence of faraway lands reachable by sea. Pieces of pumice floated across the ocean long before there were rafters to take advantage of them. Seaweeds, trees, seeds, and animal carcasses later joined the armada. The marine iguanas of the Galápagos Islands must have arrived on a raft.

More recently, man has made hitchhiking much easier by vastly increasing the amount of floating debris in the form of tar balls and especially plastics. The presence of a giant Pacific garbage patch has drawn much attention, and where this debris accumulates, it is far more abundant than plankton and fills the guts of hapless seabirds. But even when Thor Heyerdahl himself rafted across the Pacific on the *Kon Tiki* in the 1940s, the most common floating material he encountered was plastic. By the 1980s, ships released eight million pieces of debris every day.

Some animals have made hitchhiking, particularly on other animals, a permanent way of life. There are nearly 30 species of barnacles that are found only on turtles, and another group that are found only on whales. One group of shrimplike creatures specializes on living with gelatinous plankton such as jellyfish and comb jellies. Perhaps one day, species that specialize on plastic will evolve. For these perpetual travelers, it's all about the journey, not the destination.

▲ A shrimplike amphipod holds on for a free ride, sailing the seas on top of a jellyfish (*Mitrocoma cellularia*).

FAST FACT:
Giant ocean whirlpools called gyres collect floating objects. The Atlantic has its Sargasso Sea, named after the seaweed sargassum. The seaweed eventually decays and provides food in the process. The Pacific has its Great Garbage Patch, which is twice the size of the United States. The plastic also degrades, but instead of food, it produces dangerous toxins.

◀ All aboard! *Chelonibia* barnacles cruise exclusively on green turtles (*Chelonia mydas*) as their only mode of transport.

REINVENTING THE WHEEL

▲ Just as a rolling stone gathers no moss, this rolling coral *(Porites lutea)* gathers no seaweed as it tumbles across the seafloor.

FAST FACT:
Swimmers may evolve new ways to get around, and even nonswimmers can find ways to become waterborne. Dumbo octopuses flap their "ears," scallops clap their shells together, sea slugs and flat worms do a version of the butterfly stroke, and one sea cucumber creeps along at less than an inch per minute before puffing up and swimming away.

WHEELS ARE A DOMINANT FEATURE OF MODERN transportation systems, and they can be highly efficient. Under ideal conditions, it is actually easier to use a wheelchair than it is to walk. One might then think that wheels would be common in the natural world as well. But wheels and other forms of rolling locomotion are actually fairly rare among animals and plants.

Tumbleweeds are the epitome of passive rollers on land, and a few spherical corals employ the same strategy under water—the proverbial rolling stones that gather no moss. When it comes to more active rolling on land, a spider, the pangolin (or scaly anteater), and several imaginary creatures have this trick in their repertoire. Under water, they are joined by a remarkable mantis shrimp that back-somersaults across sandy beaches, a feat otherwise largely restricted to circus acrobats and gymnasts.

The long and thin *Nannosquilla* swims by choice if there is any water, but its legs are not strong enough for walking. So, if it is stranded on moist sand, it just plants its tail above its head and then pulls its body around. Although none too speedy (about one-tenth of a mile an hour), it can climb inclines of up to 30 degrees and is capable of performing as many as 40 consecutive somersaults and covering a distance of about six feet, which usually brings it back to the water.

Could natural selection be simply unable to invent wheels on a regular basis? Perhaps, but closer consideration suggests that wheels excel only under rather particular circumstances, and that their rarity instead reflects the fact that generally speaking, legs and fins work better. Wheels do best on straight, smooth, and firm roadways, surfaces that are not very common in the absence of human construction and maintenance. So *Nannosquilla* is the proverbial exception that proves the rule—what better roadway could one ask for than a firmly packed beach at the water's edge.

▶ This sea cucumber (*Enypniastes* sp.) both crawls and swims at a depth of 9,000 feet. The pink-tentacled mouth connects to a mud-filled gut.

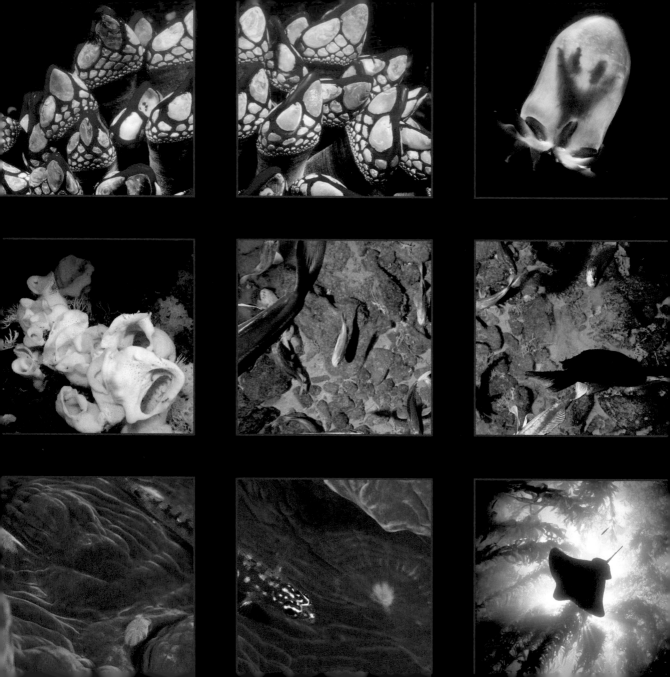

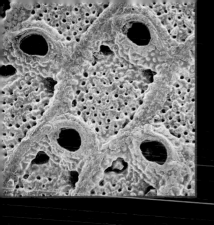  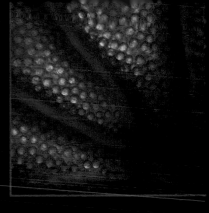

5

SETTLING DOWN & GROWING UP

People grow up and then settle down, but in the sea, the order is usually reversed. Many ocean dwellers float for a while in the plankton before heading to the seafloor. For those that then stay put, choosing where to settle is a life-and-death matter, and they use sights, smells, sounds, tastes, and textures to decide. Some animals and plants grow quickly and die young. Others, particularly those that clone, live to a ripe old age.

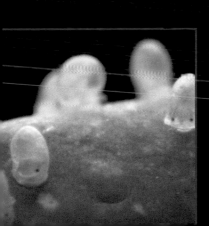 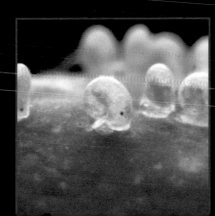 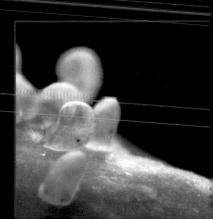

THE BIG DECISION

TO SETTLE OR NOT TO SETTLE. THE BABIES, OR LARVAE, of many marine organisms drift with the currents before they join the adult world on the seafloor. Once they are ready to leave life in the plankton, the choice of a settlement site is arguably the most important one they will ever make. Whether they are simply sedentary, like snails, or actually glued down, like barnacles or oysters, the "decision" of where to call home is often an irrevocable one.

Tiny marine animals and seaweeds often have larvae that don't travel far—they either skip the larval stage or are ready to settle almost immediately. "What's good enough for Mom is good enough for me" is their operating principle. But most marine offspring travel farther afield—some stay in the general vicinity of their parents, but others ply the high seas for as long as a year. How do they know where to put down roots when the time to settle arrives? The cues they use (and there are often several) depend on what matters most for future success.

Baby coral reef fish are attracted to reef noises, especially the popcorn-like cacophony of snapping shrimps snapping. Snails that prey on corals are looking for future food, and so they key onto chemicals from colonies that would make a good meal. Anemonefish need to find sea anemones, so they are attracted not only to them but also to the scent of leaves, because sea anemones are especially common near shores with overhanging branches. Some sea urchins and snails are attracted by fleshy or stony seaweeds, or the film of bacteria, or biofilm, that grows on them. Barnacle babies are especially attracted to biofilms from places where they are likely to prosper. Still other animals go for the most direct signs of success, and settle near or even on adults of their own species.

A complicated brain isn't needed to respond to a cue and find a nice home. But whether on land or under water, when it comes to real

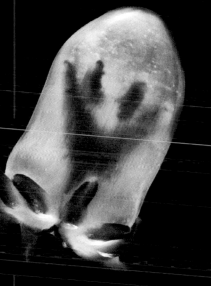

FEATURED: *Snails, Barnacles, Burrowing Anemones*

▲ Once it has settled, this burrowing anemone will build a mucus tube into the ocean floor, allowing it to hide at a moment's notice.

FAST FACT:
Some larvae put off settling down for a long time. Snail babies of one species stayed afloat in the lab for 55 months—in the ocean, they could have traveled more than 8,000 miles in that time. Though such long journeys are probably rare, a year adrift is not uncommon for lobsters and eels.

▲ Baby red abalones *(Haliotis rufescens)* prefer to settle on the crusts of stony red algae, which they recognize by using chemical cues.

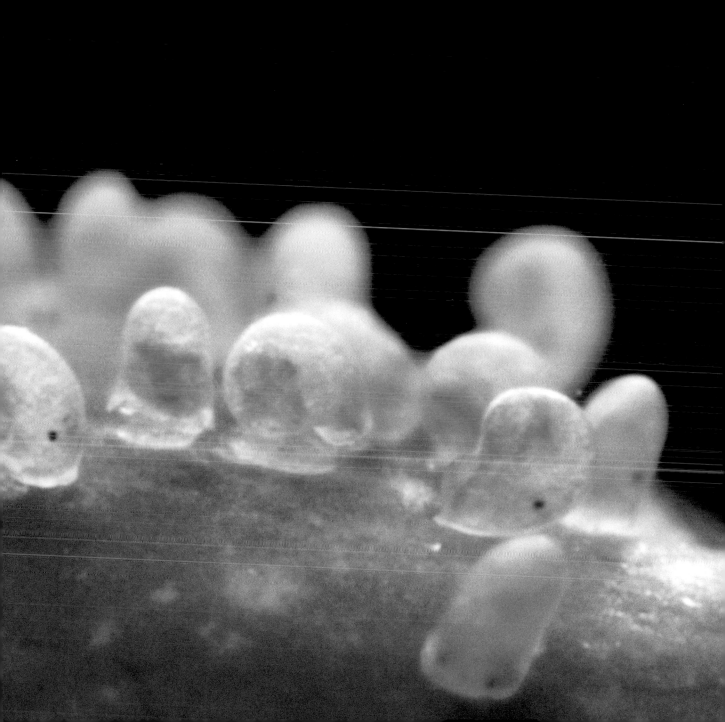

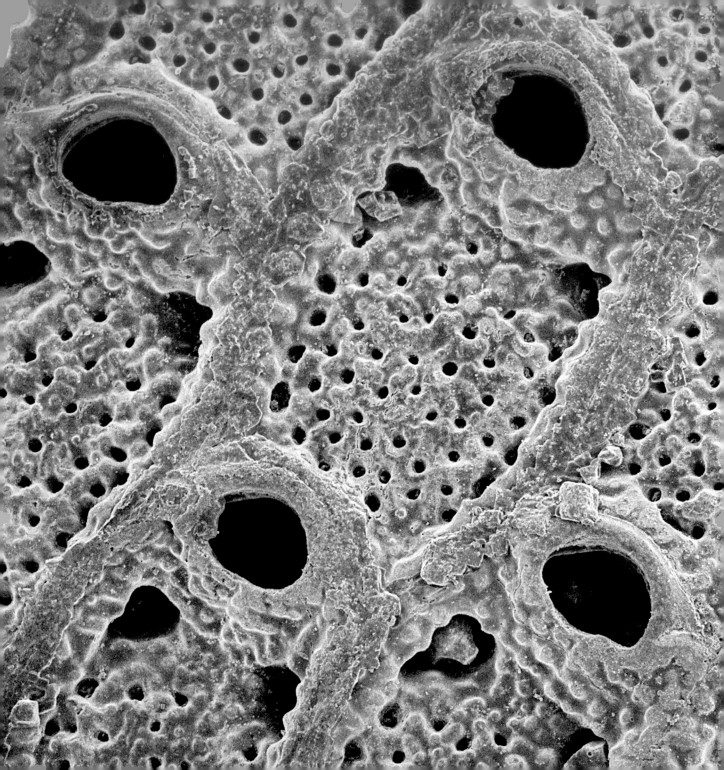

SIZE MATTERS

FEATURED: *Sponges, Bryozoans, Sand Dollars*

"N ORMAL" ORGANISMS, IF YOU DEFINE PEOPLE AND snails and insects as normal, start off as a fertilized egg that develops into a juvenile and then into a sexually mature adult that eventually ages and dies. Going backward from old age to juvenile or reproducing by splitting down the middle are not options—nor can adults grow indefinitely.

Not so for cloners. They also start as a fertilized egg, but they grow in much the same way that a brick wall is built. The basic modular unit, the brick, is standardized, but the wall, or colony, can take almost any shape or size. Many corals and moss animals can grow indefinitely by budding new "bricks," and because they are a collection of self-sufficient modules, they can survive being split into pieces. Other cloners, such as sponges, don't have an obvious bricklike structure but have such flexible body plans that they can regrow from pieces anyway—pass a sponge through a fine mesh and it will reassemble on the other side. A few cloners, such as some brittle stars, seem to have a fixed shape and adult size but are still able to regrow a new body from a broken piece. When genetically identical pieces go their own way, one has to rethink what the notion of an individual means.

Cloners often lie about their age—a large colony is definitely old, but a tiny scrap of a colony might be even older. It's size, not age, that matters, though—no matter how old, reproduction is impossible if the colony is too tiny. Sometimes modules are more like Lego blocks than bricks, varying in form and function. In moss animals, some of these "blocks" turn into egg chambers and others into defensive spines. This division of labor takes an extreme form in the Portuguese man-of-war, which looks like a normal jellyfish but is actually a colony—some modules making tentacles, others making the float.

So, while cloning is abnormal for some groups, it is conventional for others. Dolly the cloned sheep may have made headlines in 1996, but if we were built like corals, there would have been no story.

▲ This as yet unidentified sponge has been dubbed the Picasso sponge because of its deformed shape. Some colonies grow three feet tall.

FAST FACT:
Cloning not only helps you get big but it can also help you get small —and get away. Fish will eat baby sand dollars if they see them, so baby sand dollars put in water containing fish mucus can sense there might be trouble. Their solution is to divide into pieces, making themselves smaller and harder to see.

◄ When it is magnified 21 times its original size, one can clearly see the building

LIFE IN THE FAST LANE

▶ A hairfin pygmy goby *(Eviota prasites)* enjoys its short life perched on an elephant nose coral in the Red Sea.

FAST FACT:
To grow fast you need lots of energy. Animals that can use chemical energy from hot gases spewing out of the seafloor are speed demons. One worm can extend its tube almost half an inch in 24 hours, a growth spurt that represents about 40 percent of its own body weight.

TIME IS MONEY, OR IN DARWIN'S CURRENCY, OFFSPRING. Being able to grow and reproduce quickly often pays off, especially when the risks of being eaten are high, when good conditions come and go quickly, or when the winner takes all.

Giant kelps, the largest plants in the ocean, are in a race to the top. These enormous seaweeds need to reach the surface, often from depths of 100 feet or more, in order to have enough light to make the food they need to reproduce. Periodically, El Niño weather patterns bring big storms that uproot them or bring warm water that lacks nutrients and can be equally lethal. The resulting mass mortalities open up space, and newcomers that get to the surface first spread out their fronds and shade their slower neighbors. So giant kelps need a lot of energy to win this competition (which is why they do poorly when nutrients are low), sometimes growing up to two feet a day. But rapid growth comes with a price—living fast means dying young, and giant kelps are surprisingly short lived given their size, usually lasting only a few years.

Many organisms that live in the fast lane are much smaller than giant kelps. The champion speedster among animals with backbones is the coral reef pygmy goby. Smaller than an inch and with a death rate of nearly 8 percent a day, they are in a race against time. After three weeks in the plankton, they settle onto reefs and mature in about ten days. Females typically produce three clutches over the course of the next three weeks before dying. Total life span? No more than a fleeting 59 days.

Sometimes the race to reproduce is so intense that there is no time for growing up at all. The loriciferans are tiny animals, less than two-hundredths of an inch, that live between sand grains. These Russian dolls of the sea have larvae inside larvae inside larvae, with adults inside some of the larvae as well. Having young before even being born takes the concept of breeding like rabbits to a whole new level!

▶ Fast-growing giant kelps *(Macrocystis pyrifera)* form cold-water forests that many fish, invertebrates, and mammals call home.

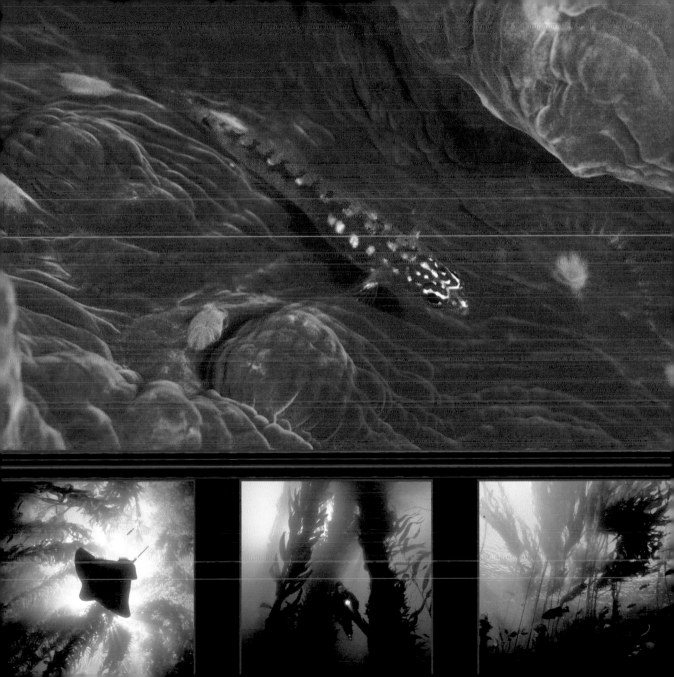

METHUSELAHS OF THE DEEP

NO ORGANISM POSSESSES THE SECRET OF ETERNAL life, but some come close. For them, slow and steady wins the race. This strategy makes sense when resources are scarce, or when restraint early on leads to big payoffs later.

In the deep sea, food is generally scarce, so growth is slow, and the option of life in the fast lane is ruled out. Even small animals (with no light there are no plants) can be surprisingly old. Orange roughies are a common fish on seamounts that only grow to about a foot and a half in length. But the rings in their ear stones (which are counted like tree rings) tell a tale of surprising longevity—they can live up to 125 years and don't even start to reproduce until they are about 20.

The oldest organisms, on land and in the ocean, are those that clone and can increase in size indefinitely. The larger they are, the more offspring they can make, so the advantages of old age can be substantial. Judging by rings in the skeleton, one colony of deep-sea black coral was estimated to be 4,265 years old, making it a contemporary of Egypt's 6th dynasty pharaohs and the oldest marine organism ever found. Because so few corals have been dated, even older corals are certain to be discovered.

These ancient mariners of the deep used to be out of our reach commercially, but no longer. Orange roughies were introduced to world markets in the late 1970s, but catches have declined sharply since their peak in the 1990s. And no wonder—with such a slow growth rate they only marginally qualify as a renewable resource. Even worse, trawling takes not only the fish but their habitat, the deep-sea corals and other animals living with them. This underwater strip-mining causes the richness of these communities to decline by a thousand fold, and because deep-sea corals grow so slowly, recovery can take decades or longer. Fortunately, countries around the world have begun to protect these rich but vulnerable habitats, where even the 969-year-old Methuselah might have been viewed as a youngster.

▲ Living for thousands of years, black corals such as these *(Gerardia savaglia)* are the senior citizens of the sea.

FAST FACT:
Yearly rings in the skeletons of long-lived corals tell us far more than just their age. Growth rates show how healthy the coral was, and the chemistry of the skeleton itself is a kind of stony thermometer and rainfall gauge, recording past monsoons and El Niños, as well as pollutants washed into the sea.

FEATURED: *Orange Roughies, Black Corals*

◀ Off the coast of New Zealand, these orange roughies *(Hoplostethus atlanticus)* live a long, slow life on their seamount homes.

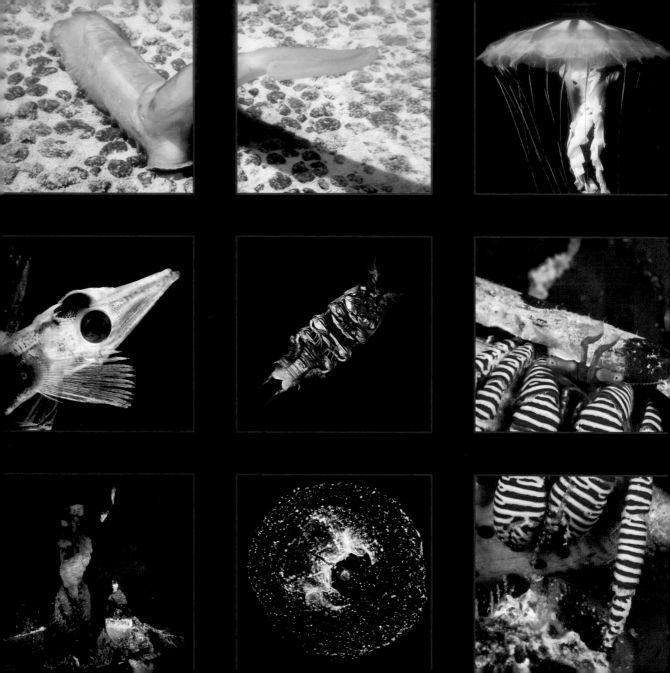

THERE'S NO PLACE LIKE HOME

Everything's relative—unbearably hot for some organisms is inhospitably cold for others. The ocean provides a nearly unlimited array of temperature, pressure, and illumination options—the icy chill of the Poles, the darkness and crushing pressures of the deep sea, and the scalding sulfurous waters of hydrothermal vents. But some organisms need a house as well. For parasites, that house is another organism.

HOT AS HADES

FEATURED: Microbes, Pompeii Worms

▲ The surface (in yellow) of this worm *(Alvinella pompejana)* is covered by a "farm" of bacteria (blue) that survive on the chemicals from hydrothermal vents.

FAST FACT:

Vents are ancient ecosystems, but individual vents are not. They are constantly being born and dying, and many have life spans of only decades. The offspring of vent animals are carried by ocean currents to new vents, and it takes only a few years for cooled rock to develop a thriving vent community.

IT CAN GET PRETTY HOT AT THE SHORE ON A CLOUDLESS day, as anyone walking on a baking beach barefoot can testify. In lagoons of the Arabian Gulf, temperatures can reach more than 100°F, and on mud- and sand flats, a staggering 150 degrees. "If you can't take the heat, get out of the kitchen," or so natural selection seems to be saying.

But the organisms that are really in hot water are those that live near hydrothermal vents. At the mid-ocean ridges where new seafloor is pushed to the surface, water seeps into cracks, is heated by molten rock, and can bubble or gush out at temperatures of more than 750 degrees. Metals and sulfur quickly come out of solution when they hit the cold surrounding seawater, forming clouds and chimneys. Paradoxically, these black smokers are oases of life, thanks to the ability of microbes to harvest the chemical energy of the vents and pass on the fruits of their labors to animals. These microbes push the limits of adaptability and live within inches of being boiled alive (figuratively speaking—because of the pressure, the water doesn't actually boil).

Pompeii worms and their relatives are the champion animal survivors here. They not just survive the heat, they prefer it. When given a choice, one species opts for water of 100–120 degrees and will even check out temperatures of 130 degrees for 15 minutes or so. But these worms live close to the edge—in water just ten degrees warmer, they die within minutes. The worms also have to deal with 100-degree fluctuations in temperature, little oxygen, corrosive acidity, and concentrations of poisons that would kill most creatures. Yet even these animals are warm-water weaklings compared with the microbes. Strain 121, the current record holder, not only survives but also reproduces at temperatures of 250°F.

Hydrothermal vents come in many forms, and are as old as the oceans themselves. Ironically, though they seem extreme to us now, they may have been the chemical cradle for the evolution of life itself.

▶ At a depth of 11,500 feet, black smokers at the mid-Atlantic Ridge spew out 660°F water laced with chemicals that support abundant life.

COLD AS ICE

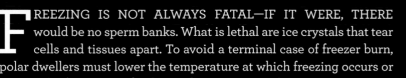

FEATURED: Icefish, Emperor Penguins, Ice Worms

FREEZING IS NOT ALWAYS FATAL—IF IT WERE, THERE would be no sperm banks. What is lethal are ice crystals that tear cells and tissues apart. To avoid a terminal case of freezer burn, polar dwellers must lower the temperature at which freezing occurs or control the formation of ice crystals to minimize damage.

The first solution is familiar to anyone that drives a car in the cold—antifreeze. The salt in seawater is itself a mild antifreeze (which is why salt is sprinkled on sidewalks in winter). Seawater typically freezes at just under 29°F rather than the 32 degrees at which freshwater turns into ice. But the blood of fish is less salty than seawater, so they need added protection. Different kinds of antifreezes are found in a variety of fish, ranging from cod to herring to flounder.

The most famous examples are the icefish—over a hundred species that represent 45 percent of the fish species in Antarctica and, in the coldest waters, more than 90 percent of fish flesh by weight. Fish antifreezes work by binding to the surfaces of ice crystals, preventing additional water molecules from joining. The most powerful antifreezes allow fish to swim ice free at 27.5°F.

Cold as this seems, it is positively tropical compared with the temperatures faced by polar marine organisms that spend time out of water. Temperatures in the intertidal can drop to minus 4°F. To survive, animals and seaweeds use chemicals that keep ice from forming inside cells and that keep cells from collapsing as water moves out of them.

Being warm-blooded is the ultimate solution for beating the cold, but keeping blood warm can be challenging. Emperor penguins need to do this both on land, where howling winds and temperatures of minus 60° may prevail, and in the sea, which though warmer, can suck heat away from an unprotected body in minutes. Not surprisingly, they have the highest density of feathers of any bird—100 per square inch—and when that fails, they huddle.

▼ A jellyfish (*Chrysaora melanaster*) gracefully hovers deep in the cold waters of the Canadian Arctic.

FAST FACT:
It's not cold enough on the deep-sea floor for seawater to freeze, but pressure can entrap methane gas in rigid cages of water molecules. These icelike methane hydrates may be the most abundant fossil fuel on Earth, and they are also home to the "ice worm," which eats bacteria that grow in the ice.

◄ Thanks to the natural antifreeze in its blood, this juvenile icefish (*Chionodraco hamatus*) survives in icy Antarctic waters.

LIFE UNDER PRESSURE

▶ More than 16,000 feet deep, this sea cucumber *(Psychropotes longicauda)* uses its "sail" to take advantage of underwater currents and glide across the seafloor.

FAST FACT:
The deep sea has traditionally been viewed as a place to fish, mine, and dump wastes. But in 2009, the Marianas Trench Marine National Monument was established. It protects not only the deepest spot on Earth but also other extreme environments sure to provide new examples of unusual ways of surviving.

THE DEEPER YOU GO IN THE OCEAN, THE GREATER THE pressure—it doubles every 33 feet. At depths of 20,000 feet, which cover more than half the Earth's surface, the pressure is nearly 600 times that at sea level. But this is still shallow, compared with the depths of the 22 trenches that score the seafloor. The deepest of these is the Mariana Trench—at more than 36,000 feet or nearly seven miles, it is much deeper than Mount Everest is high, and the pressure is more than 1,000 times that felt at the surface.

It was long thought that nothing could live in the dark, cold, and crushing pressures of the deep sea. That changed with the *Challenger* expedition in the 1870s, which brought up animals from more than 18,000 feet and mapped the trenches. In the 1950s, Danish and Russian ships sampled the trenches themselves, and in 1960, for the first and last time, men visited the deepest part of the Mariana Trench in person in the bathyscaphe *Trieste.* Since then, only two unmanned vehicles have made the trip. In 2008, this last frontier was declared a U.S. National Monument.

What lives in the deepest parts of the deep sea? Sharks are gone by about 13,000 feet, but bony fish have been trawled from more than 27,000 feet. The deepest parts of the trenches are home to shrimplike creatures, worms, sea cucumbers, and of course, microbes.

Rapid changes in pressure are hard on any organism—bug-eyed fish brought up from even modest depths are the victims of reduced pressure that causes their swim bladders, which they use to control buoyancy, to blow out. But high pressure itself requires special adjustments so that the molecular machinery of the cell continues to function. Especially vulnerable are the fats that make up the cell's membranes—the effect of high pressure is like that of low temperature, so that to animals of the Mariana Trench, the chilly 35°F feels like a frigid minus 2°. They compensate by replacing saturated with unsaturated fats, which as cooks know, are more liquid in the cold.

▶ What do a sea cucumber, copepod, and comb jelly have in common? Living at depths of 8,000, 16,000, and 3,000 feet, they can handle the pressure.

MOBILE HOMES

MANY OCEAN ANIMALS LIKE TO HAVE SOME KIND OF roof over their heads to provide protection against their predators. For some, this means a crevice in a rock, for others a burrow in the sand. But for those who need to move from place to place, something more portable is required.

Hermit crabs have solved this problem by taking up residence in the shells that snails leave behind when they die. Pick up an "empty" shell, and often it turns out to be occupied. These shelters are not optional—over time, the hermit crab's back end has evolved to become soft and vulnerable, and without a shell it gets eaten. And not just any shell will do—hermits are picky if given a choice, and they trade shells with other hermits to get a better fit. Like a cautious home buyer, they check the shell's size and look for any signs of damage.

Some homes provide not protection but food. Larvaceans are tadpole-like creatures (related to sea squirts and to us) that feed by filtering seawater. Most filter feeders use tentacles of some sort, but larvaceans build a house of mucus. Each house has two filters—a coarse outer filter and a fine inner one—and the larvacean uses its tail to pump water through them. In deep open water, the houses can be a yard across.

Larvacean houses are flimsy affairs with a limited life span. When the filters get clogged, the tadpole discards its house and builds a new one, frequently as often as once a day. Unlike the hermit crab, whose home gets passed on to other hermits until it really wears out, no other larvacean will use the discarded home. But that does not mean that it goes to waste. The large houses collapse and sink rapidly as marine "snow" at the rate of half a mile a day (making them mobile indeed). These discards provide welcome food to the hungry creatures on the seafloor below.

Which all goes to show that recycling is the rule in the ocean, and what is one animal's garbage is another's gold.

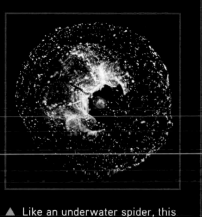

▲ Like an underwater spider, this larvacean *(Oikopleura labradorensis)* traps food in the mucus net it has built.

FAST FACT:
Some Indonesian octopuses wander around stilt-legged, with a pair of half coconut shells stacked under the body. When cover is needed, the octopus jumps in one half and pulls the other half on top for a roof. It's the first known case of tool use in a spineless sea creature.

◄ Home sweet cone! This cone shell is a perfect fit for the flat body of the hermit crab *(Ciliopagurus strigatus)*.

UNINVITED GUESTS

▼ This crab *(Liocarcinus holsatus)* plays unwilling host to a parasitic barnacle *(Sacculina carcini),* whose eggs the crab carries instead of its own.

FAST FACT:
Microbes can't be seen with the eye, but they can still make sea organisms sick. The black long-spined sea urchin of the Caribbean used to carpet reefs in shallow water. Over the course of a year in the early 1980s, 95 percent of them died in an epidemic. The culprit remains unknown.

PARASITES TURN OTHER ORGANISMS INTO THEIR HOMES. Most don't kill their hosts, at least not quickly—it can be hard to find a new home. Sometimes the host is just a shelter, but often it winds up involuntarily serving dinner (or serving as dinner) too. The consequences range from innocuous to disastrous.

The lifestyle options for parasites are almost limitless. One group of shrimplike creatures is the bane of fish, attaching to the host's forehead or, more gruesomely, eating the host's tongue to make more room in the mouth. And who would have thought that fish might view the rear end of a sea cucumber as a promising piece of real estate? Apparently, pearlfish do. Once inside, some pearlfish do comparatively little damage, but others are castrators, eating the testes and ovaries.

Fish are not the only parasites that turn their hosts into genetic dead ends. Young barnacle parasites start off looking like their cousins on the shore (which is the only reason they were recognized as barnacles in the first place). But when they leave the plankton, they ignore rocks and instead settle onto their crab or shrimp hosts. Female parasites inject a bit of protoplasm, which then grows roots that take over the host's body. Eventually, a bulging sac emerges from the host's belly, and baby male parasites enter the sac to fertilize the eggs inside. When the next generation of parasites is ready to be released, the parasite even manipulates the host's behavior so it will help launch them.

Parasites may be unwelcome (at least to their hosts) but they are impossible to avoid or ignore. In fact, their weight in a bay or estuary can exceed the weight of all the top predators combined. Every species in the ocean has its own specialized cast of tormenters, and some parasites do double duty, making a living by being passed from one species to another. Though not always in plain sight, they are quite literally everywhere.

▶ This worm *(Gastrolepidia clavigera)* crawls into the anus of its sea cucumber host. Luckily for the cucumber, the worm is much smaller than a pearlfish.

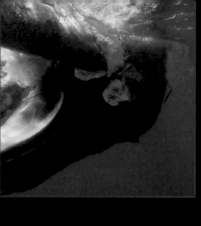

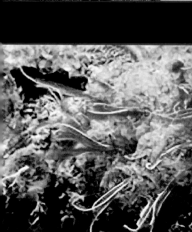

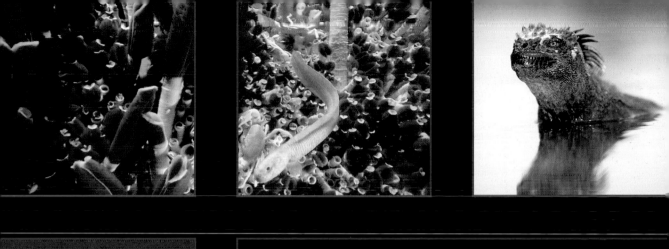

SEA FOOD

In the ocean, predators use far more than teeth to subdue their prey, and meals come in many forms. In the darkness of the deep sea, microbes make it possible to capture chemical energy from hydrothermal vents and whalebones. Even in shallow water, unusual solutions to the age-old problem of finding dinner abound. Fish farm the seafloor, and lizards stranded in the Galápagos dine on seaweeds.

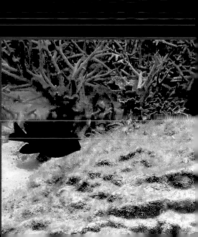

WEAPONS OF CHOICE

▲ Only six inches long, the fang-tooth *(Anoplogaster cornuta)* is a well-equipped deepwater preda-tor. Small fish that enter its fanged mouth are unlikely to exit.

FAST FACT:
Some predators resort to shocking methods to kill their prey. Upon spot-ting a potential meal, the torpedo ray lunges and wraps its disklike body around its victim in an electrifying embrace. Attacks can involve more than 1,200 jolts, each as much as 60 volts.

HUMANS, BEING THE TOP PREDATOR OF THE PLANET, now have little to fear from man-eaters in the ocean, the movie *Jaws* notwithstanding. But most marine animals are eaten (usually sooner rather than later) by an array of predators, some of whose tools of the trade make the razor-sharp teeth of the fangtooth seem positively boring.

Claws with hammers are the weapon of choice for mantis shrimps that eat hard-shelled prey, such as snails. The blow is delivered with extraordinary speed. The hammer can move more than 50 miles an hour, and the strike happens so quickly (it takes less than three-thousandths of a second) that even filming at 5,000 frames per second misses some of the action. They are, indeed, the fastest gun in the west—with a speed comparable to a .22-caliber bullet, no animal arm or leg is faster. To accomplish this feat, the mantis shrimp stores energy in elastic springs, much as a crossbow stores the energy of an archer, and like a crossbow, the strike is precisely timed by releasing a latch. The force initially delivered is thousands of times greater than the weight of the animal, and indeed is considerably greater than the force of gravity on an adult human body. Even water-vapor bubbles are formed, which amplify the damage. It is thus no surprise that big mantis shrimp can break through the glass wall of an aquarium.

On the slower but weirder side of the spectrum are the ribbon worms that prey on fiddler crabs as they walk about the sand to feed and mate. Unlike the sandworms of the sci-fi classic *Dune*, ribbon worms rely not on speed but on stealth. Lurking below the surface, the worm emerges when a crab passes overhead, turning its proboscis inside out and covering the crab with toxic slime. Once the victim is paralyzed, the worm finds a crack in the body by which to enter and crawls inside, digesting the crab from the inside within an hour. And these ribbon worms are small fry—some of their cousins reach 100 feet in length, which is as long as a blue whale.

▶ Powerful strikes from the bright red hammer of this mantis shrimp *(Odonto-dactylus scyllarus)* make quick work of the shells of its prey.

The weapons of this mantis shrimp *(Odontodactylus scyllarus)* are put to more peaceful purposes. This mom uses her clubbed claws to carry her clutch of eggs.

FISH FARMERS

FEATURED: Damselfish, Marsh Snails, Polysiphonia Algae

▲ This salt marsh snail *(Littorina irrorata)* is a fungiculturist, enjoying a homegrown meal on saltwater marsh grass in North Carolina.

FAST FACT:
People, ants, termites, and beetles farm fungi, and so do salt marsh snails. They use their teeth to open wounds on marsh grasses, which are then colonized by their favorite fungal food. The snails even speed the process along by pooping on the wounds, introducing bits of undigested fungi.

THE INVENTION OF AGRICULTURE TRANSFORMED human culture about 10,000 years ago. Simple bands of roving hunter-gathers were largely replaced by sedentary and sometimes complex societies. Humans are not alone in having invented agriculture, but it is not common in nature. On land, the most conspicuous nonhuman farmers are the new-world leafcutter ants and a group of old-world termites. Under water, damselfish farm the seafloor.

Damsel farms or gardens are conspicuous on coral reefs around the world. Here and there, the cover of living coral is interrupted by patches of seaweed turf, and in some places, gardens are actually more common than coral. Look too close, and a small but fearless fish will greet you, rising up out of the coral to deliver sharp little bites. Even when confronted by a band of hundreds of marauding parrotfish, the damsel is indefatigable in the defense of its crop. Like human farmers, the damselfish has created a tasty and tempting meal that would be consumed by others in an instant if left unattended.

The best damsel farmers weed their gardens. They bite and remove any unwanted invaders that would otherwise quickly take over, maintaining small, dense patches of highly nutritious monocultures. Other damselfish are less committed farmers, rarely weeding their gardens, which are larger and more diverse but less nutritious.

Plants intensively cultivated by people changed as they were domesticated and became dependent on people. The same is apparently true of the seaweeds farmed by the most committed damselfish. These damsel domesticates are much less diverse than human crops, all belonging to a group of seaweeds called *Polysiphonia*, but like domesticated crops, the *Polysiphonia* that dominate damsel gardens are rarely or never found growing wild. And just as some farmers grow wheat and others corn, different species of damselfish grow different species of *Polysiphonia*. All the damsel farmers seem to be missing are tractors, electric fences, and markets.

OCEANGOING IGUANAS

THE GALÁPAGOS ISLANDS LIE MORE THAN 600 MILES from the coast of Ecuador, and they have never been attached to land. The ancestor of the three species of iguanas that live there must have hitched a ride to these volcanic islands on some floating debris. Their DNA suggests that they arrived about 11 million years ago, before the current islands appeared, landing on a piece of rock that has since sunk below the surface.

These immigrants were presumably terrestrial at the outset, but at some point some ventured onto the shore to look for food. Thus began a dietary odyssey that led to the evolution of the Galápagos marine iguana, the only species of lizard that routinely goes to sea and eats primarily seaweed. The Galápagos get little rain, and the land iguanas there mostly subsist on cactus, so you would think that seaweeds would be an easy food source. But for a cold-blooded lizard, diving into the chilly water of the Galápagos poses problems, and most of the seaweeds are covered by water except during low tides. As a result, marine iguanas feed for only about two hours a day, and the rest of the time they spend basking in the sun to warm up. Only the large males are able to dive down to reach submerged seaweeds—females and juveniles are limited to what is uncovered when the water recedes.

The El Niño climate cycle brings additional challenges. Torrential rains bring the land to life, and the two species of land iguanas enjoy the bounty, but the warmer sea temperatures cause most of the nutritious seaweeds to die off. As much as 90 percent of the marine iguanas may die in a severe El Niño. The ones that survive shrink as much as 20 percent in body length—an ability unique among vertebrates.

Darwin described marine iguanas as disgusting and clumsy, and he named them the "imps of darkness." But their black skin, webbed feet, and gripping claws, which allow them to absorb heat, swim, and grip the sea bottom, make them a perfect example of the evolutionary process that Darwin went on to describe.

▲ The waters of the Galápagos are chilly, so marine iguanas (*Amblyrhynchus cristatus*) spend a lot of time basking in the hot sun.

FAST FACT:
There are no fossils of the first iguanas to arrive in the Galápagos, but molecular "clocks" can tell us roughly when they got there. Some mutations in genes accumulate fairly steadily, so DNA differences between today's marine iguanas and their mainland relatives are a measure of how long ago marine iguanas went to sea.

WHALE FALL WINDFALLS

▶ A young humpback whale (*Megaptera novaeangliae*) meets an early death. Its carcass will fall to the ocean floor and become food for hundreds of species.

FAST FACT:

Two new deep-sea snails were recently discovered, dining on whale-bones off the coast of California. They look like fossils of snails from the time of the dinosaurs, but whales did not evolve until after the dinosaurs went extinct. Back then, huge seagoing reptiles, the plesiosaurs, were swimming, dying, and falling to the seafloor, providing food for the bone-eaters.

IN THE LIGHTLESS DEEP SEA, THERE ARE NO PLANTS AND hence no vegetarians. Most animals are dependent on the little food that comes from above, usually as a gentle rain of small particles. But occasionally a windfall arrives in the form of an entire carcass. When that carcass is a 160-ton whale, it feeds deep-sea dwellers for decades. Scientists in a submersible found their first whale carcass in 1987, and since then they have found (and sunk) more, allowing them to study what happens when a dead whale hits the bottom.

The first to arrive at the scene are the hyenas and vultures of the deep. Hagfish (primitive jawless fish), sleeper sharks, and rattails tear at the corpse, removing more than 100 pounds of meat and blubber daily. Finally, after about two years, just scraps remain. Those scraps then feed smaller scavengers—crabs, worms, snails, and shrimps—that pick the skeleton clean, much as corpse beetles do on land. But even bones are potential food sources. Their fatty marrow is fed on by bacteria that need no oxygen and produce hydrogen sulfide as a byproduct. This gas, with its rotten egg smell, is highly toxic to most animals, but some animals have formed partnerships with other bacteria that can use it. One such worm—without mouth, stomach, or eyes—grows roots, which house bacteria that turn the bones into the equivalent of Swiss cheese. The smaller scavengers and marrow eaters reach huge numbers, with more than 40,000 individuals feeding on the remains. In the end, even that food is gone, and what bones are left become perches for animals that need solid footing.

Whale falls have more species (more than 400) than any other hard structures at these depths, and at least 20 species are whale-fall specialists. Today, at any one time, there are probably more than 800,000 large whale carcasses on the ocean floor in some stage of decomposition. But whaling has decimated the big whales, so some of the creatures that used to depend on them may have gone extinct before scientists had a chance to study them.

▶ The bare skeleton of a gray whale may not seem like much sustenance, but it can feed an army of bone-eating worms (*Osedax* sp.).

FOOD WITHOUT SUN

BOB BALLARD FOUND THE WRECK OF THE *TITANIC*, AND you would think that would be the proudest moment of his long career of ocean exploration. But ask him what his most *important* discovery was, and his answer might surprise you—a bunch of worms. Of course, not ordinary worms, but bright-red six-foot gutless wonders whose energy comes from the crust of the Earth. As Ballard puts it, "The *Titanic* we knew about. We did not know about this system." The system was hydrothermal vents and the animals and microbes that depend on them. Amazingly, plumes of hot water laced with toxic chemicals that spew out where new seafloor is being made support one of the most productive ecosystems on the planet.

Before the discovery of hydrothermal vents, all complex life was thought to be fed by sun-driven photosynthesis. Even life on the deep seafloor was supposed to get its energy from what rained down from the sunlit waters above. But in 1977, Ballard and colleagues returned in the submersible *Alvin* to the place where an unmanned vehicle near the Galápagos had spotted hot water and strange creatures, and they saw firsthand not only the giant worms but also clams, crabs, shrimps, snails, and other creatures in profusion.

How do so many animals make a living without drawing on the sun's energy? The answer is symbiosis—an intimate partnership with metabolically talented bacteria that use the energy stored in chemicals from the vent water to turn carbon dioxide into sugars, much as light energy does in plants. The giant tube worms keep their bacterial farms in a special internal organ, while clams use their gills, and other vent animals grow bacteria on their surfaces.

The discovery of these vents opened our eyes to an unexpected world, one independent of food from photosynthesis. But although the bacteria may get all their energy from the chemicals they are bathed in, the worms and crabs still need oxygen. That they do owe to the sun and the plants above that make oxygen for the whole ocean.

FAST FACT:
Apart from vents and whale falls, the deep sea has little food to go around, and some drastic changes in diet have evolved. Some species have given up filtering food from seawater to become predators. There are sea squirts that act like submarine Venus flytraps and sponges that trap prey like living flypaper.

◀ Tube worms (*Riftia pachyptila*) and eelpout fish (*Thermarces cerberus*) are among the many creatures that survive in communities supported by hydrothermal vents.

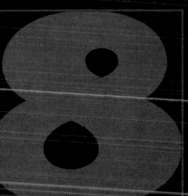

8

BEST FRIENDS

"Nature, red in tooth and claw" tells only part of the story. Although predators and competitors abound, so do relationships that are mutually beneficial. Corals partner with photosynthetic algae, fish and shrimps share burrows, and sponges intertwine. There are even doctors and dentists that remove parasites from their fishy clients. But like many a human friendship, there can be conflicts as well.

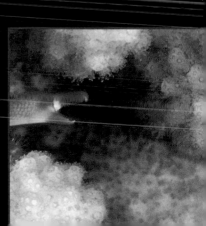

▲ This snapping shrimp *(Alpheus randalli)* and goby *(Amblyeleotris yanoi)* make a good team. The shrimp shovels sand and the fish warns of danger.

FAST FACT:
Red groupers are the most important fishery in the Gulf of Mexico, but they do much more than feed us. As teenagers, they clear out sand and mud from large sinkholes, each about 6 feet deep and 20 feet across. A 3-D seafloor is good for diversity, providing safe haven for many other species.

INNKEEPER WORMS GET THEIR NAME FROM THE FACT that their muddy U-shaped burrows are full of guests. A clam, a crab, a fish, and another worm all regularly take advantage of the innkeeper's hospitality. But unlike a true innkeeper, the innkeeper worm benefits little from the animals that share the burrow (and may even suffer a bit if they mooch too much food). In the ocean, many organisms live together to the serious detriment of none but with the benefits unevenly divided.

But one-sided profiteering does not describe the 120 species of gobies and 20 species of shrimps that share burrows throughout the world. These partners can only be described as ideal roommates. A pair of shrimps does the digging, constructing, and maintaining of the shared domicile, while the fish is the early warning system, letting the shrimps know if danger lurks. They keep in constant contact, literally, with one antenna of the shrimp always touching the goby. A flick of the goby's tail says danger is near, and the nearer the danger, the greater the number of flicks. Too close an approach and the goby dives head-first into the burrow, followed in short order by the shrimp. Gobies without shrimp burrows get eaten by predators. For the shrimp, life without a goby means that venturing beyond the burrow entrance for food is too risky, so it has to subsist on the meager rations of whatever can be found in its subterranean tunnels.

In some cases, one roommate might be better described as the room. Branching corals in the Pacific host an entire community of fish, shrimps, and crabs that depend on the coral staying happy and healthy—it provides both food and shelter. In the other direction, nutrients from fish poop and pee help the coral grow, and the shrimps and crabs sweep away dirt and attack marauding crown-of-thorns starfish, a coral predator that can eat them out of house and home in a matter of hours. They pinch the starfish's sensitive tube feet as if their lives depended on it—which of course they do.

▶ "Not MY house!" A guard crab *(Trapezia tigrina)* pinches a predatory crown-of-thorns starfish *(Acanthaster planci)* to protect its coral home.

With claws splayed, this guard crab (*Trapezia intermedia*) is ready to protect its cauliflower coral (*Pocillopora meandrina*) from either predators or intruders.

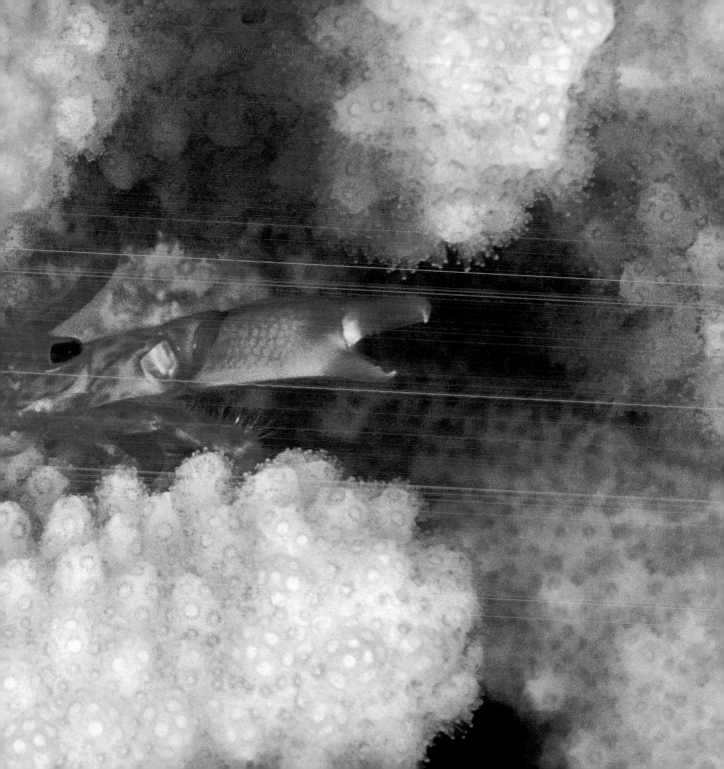

OPEN WIDE

FEATURED: *Cleaner Shrimps, Cleaner Wrasses*

PARASITES ARE AN UNPLEASANT FACT OF LIFE. OFTEN the hapless host has no recourse, but on coral reefs, more than 100 cleaner species make a living by removing and eating the tormenters. For many of these fish and shrimps, cleaning is a sideline, but for some, it is a full-time job. These doctors and dentists of the sea set up small cleaning stations, often dancing or (in the case of shrimp) waving their antennae to advertise their presence. Put a hand in front of them, and they will often do a quick exam of your cuticles.

Cleaning stations are busy places, with as many as 2,000 clients a day opening their mouths, gills, and body surfaces to inspection and parasite removal. The result is a mutually beneficial relationship, or mutualism. The cleaners get food and the clients receive health benefits. In one study, the number of parasites on fish increased four-fold after just 12 hours of cleaner deprivation.

Unfortunately, there are scam artists that take advantage of these arrangements. For the cleaners, the risk of being eaten by one of their patients is small—fish don't swallow their cleaners after enjoying their services, because the long-term need for parasite removal outweighs any short-term nutritional benefit from eating the tiny care providers, which would quickly become scarce if they were routinely snapped up. Rather, the real cheaters in the system are the cleaners themselves, a number of whom prefer fish mucus to parasites and will take small bites from time to time. But their clients notice these infractions and punish the cheaters by chasing them, swimming away, and boycotting their stations. Fish even take note of the quality of care being provided to those ahead of them in line, avoiding the cleaners that are not well behaved. Other species get into the cheating game, hanging out with and mimicking the distinctive colors of cleaners and biting unsuspecting visitors to the station.

And so nature comes full circle, with fish being plagued not only by parasites but also by fake doctors and dentists.

▲ Say aaaaah! A cleaner wrasse (*Labroides dimidiatus*) inspects and cleans the mouth of a moray eel (*Gymnothorax javanicus*).

FAST FACT:
Mom-and-pop fish-cleaning stations often give superior service. If the male catches the female stealing a bit of mucus instead of cleaning, he chases her, and this punishment makes her less likely to cheat again. Though the male cleaner was not the initial victim, he benefits in the long run by keeping his customers satisfied.

◀ It may look dangerous, but it is unlikely that this tomato grouper (*Cephalopholis sonnerati*) will harm the cleaner shrimp (*Lysmata amboinensis*) in its mouth.

ALTERNATIVE ENERGY

▼ Give a slug *(Elysia chlorotica)* some seaweed, and it will eat for a day. Teach a slug to photosynthesize, and it will eat for months.

FAST FACT:
Most of the corals' zoox look pretty much alike, but their superficial similarity is misleading. What used to be thought of as one species is a diverse collection of genetically different forms. Some of these zoox are much more heat resistant than others. This diversity may buy reefs some time in their fight to withstand global warming.

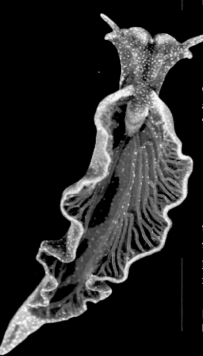

ICHENS ARE NOT ONE ORGANISM BUT TWO—A FUNGUS and a photosynthetic partner that turns sunlight into food. Beneath the waves, it's animals, rather than fungi, that have capitalized on these alternative green energy sources.

Coral reefs are the most spectacular examples. Part animal, part vegetable (and part mineral), these spectacular constructions, visible from space, are possible thanks to single-celled algae, called zooxanthellae, or "zoox" for short, that populate coral tissues. Zoox turn over much of the food they make to the corals, which allows them to grow much faster than corals without zoox can. In turn, corals provide zoox with a safe place to live and the nutrients they need to grow. Some sponges and giant clams harbor zoox as well. These partnerships are one of the great success stories of interkingdom cooperation.

From the coral's perspective, zoox that suck up nutrients but give back nothing in exchange are a losing proposition and need to be expelled. Normally this policing strategy works well, but global warming has put the coral-zoox partnership at risk. Zoox stop making food when conditions aren't right, and they are especially sensitive to high temperatures—just two degrees above normal can cause problems. Since the 1980s, warm water has caused corals around the world to kick out their zoox. Without zoox, the corals are a "bleached" ghostly white, and if temperatures don't fall quickly, they starve to death. Mortality rates of 20 percent or more can result, and future warming may make this kind of catastrophe an annual event.

As successful as the corals are, the ability to be part animal and part plant requires some fancy physiological footwork, and animal-algal partnerships are not all that easy to achieve. So some sea slugs that eat algae have taken a short cut. Instead of digesting everything, the slugs save the algae's chloroplasts (where photosynthesis takes place) in their tissues, where they provide the slugs with food for months. That's called having your cake and eating it too.

▶ When the ocean gets too warm, many corals, like this one, become "bleached" because the single-celled algae that give corals their color are lost.

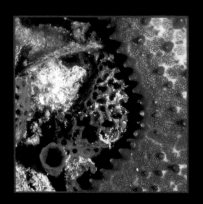

▲ Invisible from above, the stomach of this sea star *(Oreaster reticulatus)* is turned inside out to digest the tasty sponge *(Lissodendoryx colombiensis)* that lies beneath.

FAST FACT:
Sponges grow on top of mangrove roots, but they are not competitors. Sponges protect the roots from attacks by shrimp-like creatures. Mangrove rootlets allow nitrogen to pass from sponge to tree and carbon to pass from tree to sponge. Both the trees and the sponges grow better as a result.

ON THE SEAFLOOR, SPACE IS AT A PREMIUM, AND COMpetition can be ruthless. Border wars are common, and the winner is the one that comes out on top by virtue of growing faster, sometimes abetted by strategic use of poison. Yet with sponges, the outcome can be more ambiguous. Different species are often found intertwined, with no clear winner—indeed, sometimes two species are only found growing together, suggesting that they need each other. What is going on?

Sponges are fearsome competitors for space, but that is not the only thing they care about. Where storms are frequent, they need to resist the effects of waves, and where predators abound, they have to avoid being eaten. There is no supersponge, because it's impossible to be good at everything, so species that are fast growers may be fragile or tasty or both, and vice versa. On Caribbean sea grass meadows, large orange sea stars are voracious sponge eaters, and usually the only sponge survivors are the ones that taste bad. But one tasty sponge has invaded sea stars' habitat by letting distasteful sponges grow over it. It survives being smothered by growing snorkels that poke through and keep it connected with the outside world. On reefs nearby, three other sponges are often found growing together. Each has strengths and weaknesses when it comes to predation by fish and sea stars, smothering by sand and mud, resisting or recovering from damaging waves, and vulnerability to disease. By growing together, they survived a sponge holocaust that decimated many other species—a better demonstration of "if you can't beat 'em, join 'em" would be hard to find.

Sponges are not the only friendly competitors. Groupers and moray eels both hunt smaller fish—groupers catch the fish swimming and eels when the fish hide. A frustrated grouper that has chased its prey into a hole sometimes swims over to a neighboring moray and shakes its head to get the eel's help. Sometimes the eel gets dinner and sometimes the grouper, but in the long run, cooperation pays off.

▶ The complementary hunting methods of this grouper *(Mycteroperca venenosa)* and moray eel *(Gymnothorax funebris)* make them perfect partners.

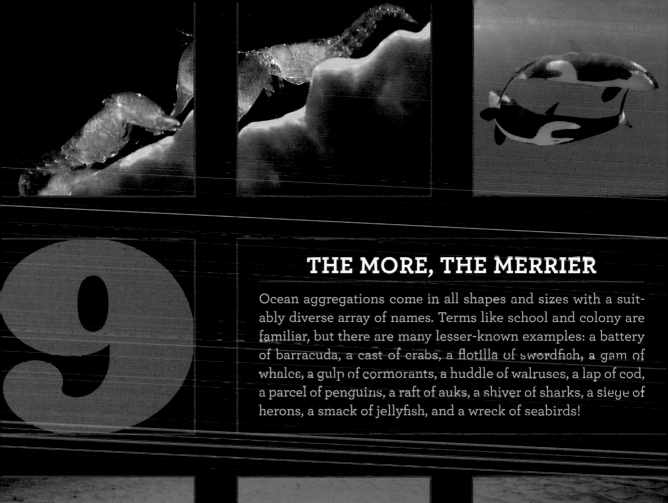

THE MORE, THE MERRIER

Ocean aggregations come in all shapes and sizes with a suitably diverse array of names. Terms like school and colony are familiar, but there are many lesser-known examples: a battery of barracuda, a cast of crabs, a flotilla of swordfish, a gam of whales, a gulp of cormorants, a huddle of walruses, a lap of cod, a parcel of penguins, a raft of auks, a shiver of sharks, a siege of herons, a smack of jellyfish, and a wreck of seabirds!

PACKS OF PREDATORS

S OME PREDATORS ARE LONELY HUNTERS, SOLITARILY stalking and chasing down their prey or waiting in ambush for them to venture too close. But others form groups to increase their chances of success. Among the most feared by fish, seabirds, seals, and whales alike are the killer whales, or orcas.

Orcas are actually dolphins, not whales, and they are found throughout the world's oceans. There are several types of orcas that differ in size, coloration, social groupings, habitat, and preferred prey, and some of these may actually be distinct species. Most orcas live in groups called pods, often consisting of females and their offspring. These groups hunt cooperatively—to devastating effect.

Some of the most sophisticated strategies are employed by orcas that prey on other dolphins, whales, and seals. Orcas off the coast of Argentina corral juvenile seals and sea lions as they attempt to enter the water. In Antarctica, orcas stick their heads out of the water to look for potential meals, spotting seals and sea lions on floating pieces of ice. The orcas then charge the ice, moving it to more open water and breaking it into smaller pieces. Finally, they swim together to create a wave that washes the victims into the sea—and into the orcas' waiting jaws. Sometimes the orcas let the prey escape several times before finishing them off with a final fatal wave, perhaps so the juveniles can learn the technique. Even sperm whales are not immune from orca attacks. A large group of orcas may break into teams, which take turns ramming and biting the whales for several hours, progressively wounding them with each attack until they succumb.

Hunting in groups has obvious benefits, but there are also costs. After the hunt, orcas often share their meals, like a pride of lions. This means that as the size of the group increases, the benefits of increased hunting success are eventually outweighed by the fact that at the end there is not enough food to go around. So the more, the merrier holds up only to a point.

▲ Even on land, this sea lion is not safe from a determined male killer whale (*Orcinus orca*), willing to beach itself for dinner.

FAST FACT:
Although they are small, territorial damselfish are probably ounce for ounce the most aggressive fish on the reef, defending their algal gardens and their developing embryos against all comers. The only way to overwhelm a determined damsel is to gang up against it—numbers eventually prevail. The bigger the attacking group, the larger the losses.

THE SELFISH HERD

FAST FACT:
The coordinated action of a human group often requires a leader, but in nature, leaders are much less common. The behavior of fish schools is but one example of how complicated patterns can emerge from simple rules. Colonial animals manage the same trick, using simple budding rules to produce complex growth forms.

WHAT'S A HELPLESS HERRING TO DO IN THE FACE OF a bunch of fast, smart, and hungry killer whales. The answer is to join forces with other herring, for the benefits of group living are not limited to predators.

Many species of small fish live in schools to avoid being eaten. This is in part simply a matter of odds—for each fish, the chances of its winding up in the mouth of a predator get smaller as the school gets larger. Also, the flashing silvery sides of many small fish make it difficult for a predator to home in on any one, and all those eyes make it less likely that a lurking predator can approach unseen. Finally, a fish can even further reduce its chances of being eaten by swimming toward the center of the school and avoiding the edges. Of course, every other fish is trying to do the same thing, so that schooling fish are constantly and selfishly maneuvering to hide behind their schoolmates.

Anyone who has watched a school of fish trying to avoid predators, or has even simply swum toward or into a school, is amazed by the seemingly coordinated way in which a school twists and turns. This coordination persists even in very large schools, numbering thousands. Schools of herring being attacked by killer whales bunch into tight balls, or smoothly split in two and then rejoin. Yet there is no head honcho herring issuing instructions. Somehow, this apparently coordinated behavior results from many individual actions guided by rules of thumb that even a cognitively limited herring can manage. Remarkably, rules as simple as 1) swim in the opposite direction of the nearest predator and 2) swim toward the nearest schoolmate until there is a risk of collision can lead to a precise choreography for the group without any leader at all.

Fish are not the only vulnerable sea creatures that seek strength in numbers—in the Arctic, thousands of juvenile king crabs can sometimes be found clustered together in tangled balls. For many organisms, being in a school can be a very smart strategy.

UNITED WE STAND

FEATURED: *Spiny Lobsters, Sperm Whales*

WHEN AMERICAN PIONEERS CROSSED THE PLAINS, they did so in wagon trains that trailed across the landscape, but when attacked, they grouped together for defense. Every fall, Caribbean spiny lobsters make their own perilous journey using a remarkably similar strategy. They walk single file across more than 100 miles of sandy seafloor to reach deeper waters. Their queues may contain as many as 65 lobsters, and when predators appear, they circle their wagons.

Lobster queues are made up of unrelated individuals, so each lobster is in it for him- or herself. The queue provides protection in numbers, and its streamlined contour also means that less energy is needed to cover what, for a lobster, is a long distance. A single-file line is not ideal, however, when it comes to defending against predators—the links between lobsters can be easily broken, leaving an isolated lobster exposed on all sides. A line also fails to take full advantage of the spiny lobsters' most potent weapons, their sharp and strong antennae. So when a predator appears (or when the tired lobsters need to rest) the lobster leader stops moving forward and instead takes a sharp turn and starts walking in a tight circle. Following the leader, the entire lobster chain slowly spirals into a giant defensive rosette, their tasty tails tucked inside with antennae lashing out in all directions. Lobsters in a rosette are much less likely to be eaten by a triggerfish, one of their most dangerous enemies, than are lobsters on their own.

Defensive rosettes are not unique to lobsters. Whales also use a circle defense with a slightly different twist—heads in and tails out.

◀ For these sperm whales (*Physeter macrocephalus*), socializing isn't just for fun. They rely on the group for protection from hungry killer whales.

FAST FACT:
When spiny lobsters are not on the move, they usually hang out in communal dens. This coziness makes it easy for diseases to spread, but the lobsters seem to know when a den mate is coming down with something serious. Even before the lobster gets visibly sick, it is avoided "like the plague."

QUEEN BEE OF THE SEA

▼ These social snapping shrimps *(Synalpheus brooksi)* have several reproductive "queens" per colony, each carrying a small clutch of eggs.

FAST FACT:
Social snapping shrimps aren't the only noisy members of their clan—their more numerous solitary and paired brethren actually make most of the underwater din. The shrimps' snaps, crackles, and pops were even the subject of classified military research during World War II because the noise was making it hard to hear enemy submarines.

UMAN BEINGS SEEM LIKE A LOGICAL CANDIDATE FOR the category of most social species on land. Like honeybees, human societies are characterized by groups in which different generations live together, and adults cooperate to care for the young. Division of labor in human societies, with their butchers and bakers and candlestick makers, not to mention TV repairmen and marriage counselors, is unmatched in the animal kingdom. Yet in one crucial way, the societies of bees (as well as ants, termites, and even naked mole rats) are more unusual. Human queens and kings, powerful though they might be, do not have a monopoly when it comes to reproduction. Insect societies might have fewer different kinds of workers—typically foragers, nursemaids, and soldiers—than do human societies, but their queen bee is the only female that produces eggs that grow up to form the next generation.

Are there queen bees in the sea? Until recently, no comparable underwater examples were known, but that changed with the discovery of large colonies of snapping shrimps living inside the canals of certain species of sponges. These might have been dismissed as simple aggregations of individuals that share a common preference for sponges, except for one key fact—even in groups numbering hundreds of shrimps, there was never more than one reproductive female, easily recognized by her near-bursting ovaries or clutches of eggs glued to her abdomen.

It is hard to explain why giving up reproduction to help raise the offspring of others evolves, but two features seem critical. First, group defense can have advantages. In the case of the social snapping shrimps, defenders snap in synchrony at intruders—creating a cacophony that deters all but the most determined—and any that persist are quickly dismembered. Second, siblings can benefit by giving up the right to reproduce if it helps the family line and their genes live on.

▶ Two of a colony's many nonreproductive adult snapping shrimps *(Synalpheus regalis)* check each other out. Their massive claws are used to repel intruders.

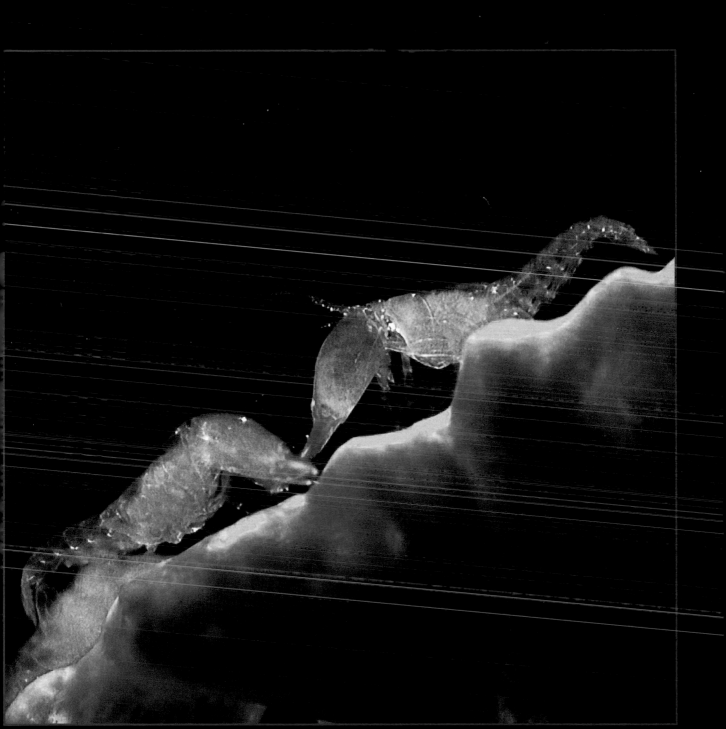

SEX & THE SEA

In the ocean, reproduction can be a decidedly unromantic affair. Many sea creatures spawn on cue and in sync with dozens, hundreds, or even thousands of others. The ocean also has its one-night stands on the beach, elaborate courtship rituals, doting parents, parasitic males, and sex changers that go from male to female or female to male, sometimes every night. When it comes to sex and the sea, there is no normal.

10

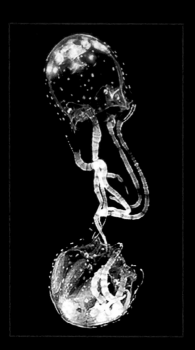

▲ These box jellyfish *(Copula sivick-isi)* intertwine their tentacles before mating, giving new meaning to "tying the knot."

FAST FACT:
Blue whale males sing to attract females. When whale numbers were down and potential mates far apart, males had to use higher notes, which travel farther. With blue whales staging a recovery, they're now singing in a lower key—the energy saved allows them to impress females (and intimidate rivals) by singing longer and louder.

BIRDS DO IT, BEES DO IT, EVEN SOME BOX JELLIES DO IT, with amorous pairs of the well-named *Copula sivickisi* entangling their tentacles before sperm is passed from male to female. Why do animals bother with courtship? The answer is simple—to persuade potential mates that they are worthy suitors. Females often need some convincing, because their nutrient-rich eggs are much more expensive to produce than sperm and they can't afford to waste them. When males spend a lot of time caring for offspring, they can be picky maters too.

A Pacific walrus male sings everything but his heart out to attract females, using his throat, teeth, nose, lips, and tongue, as well as striking his flippers against his chest. The serenades of males consist of an extraordinary array of whistles, grunts, moans, knocks, gongs, barks, boings, and burps—they have been compared with a circus at a construction site and can be heard as far as ten miles away. Females hang out in groups on the ice, and the singing males surround them in the water. Eventually, a female takes the plunge and chooses the father of her future offspring, which will depend on her for several years. Genes are all that the father will contribute, so good ones, as measured by singing style and endurance, are the best measure she has.

The situation is entirely different with blue-footed boobies. Males and females together spend more than five months incubating eggs and feeding chicks, so wooing is a two-way affair. Instead of mesmerizing their mates with melodies, boobies of both sexes engage in elaborate courtship dances that feature fancy footwork. Birds with turquoise feet are much more desirable than birds with dull blue feet, and this preference is far from arbitrary. A good supply of carotenoids (vitamin A is a familiar example) is responsible for both a healthy immune system and the sexy color—so choosy females wind up with males that are better providers, and choosy males wind up with females that lay better eggs. Who knew that being a foot fetishist could be so useful?

▶ "May I have this dance?" Blue-footed boobies *(Sula nebouxii)* show off the brightness of their feet in an elaborate courtship display.

BULLIES AND PARASITES

FEATURED: *Elephant Seals, Green Spoon Worms, Anglerfish, Barnacles*

COURTSHIP IS NOT THE ONLY WAY MALES MAXIMIZE the number of offspring they sire. In some species, males are bullies, keeping competitors away from as many females as possible. In other species, males simply race to find a female and latch on. In order to succeed, bullies must be big, but parasites benefit by being small, so they don't weaken the female too much.

The champion bullies in the ocean are the elephant seals. Male southern elephant seals average more than 7,000 pounds and are about seven times as heavy as the female. During the breeding season, males come ashore first and fight to control a section of beach—the reward can be a harem of hundreds of females. Although most contests between males are settled by bellowing (the huge nose of the male comes in handy here), injuries are also common. When the females arrive, they quickly give birth and nurse their pups for about three weeks. Females are regularly harassed and herded by the males, but come into heat only during the last few days of nursing, when males try to mate with as many as possible. Not surprisingly, pups are sometimes trampled to death in the resulting sexual mayhem.

At the other end of the spectrum are the sexual parasites—dwarf males that exchange sperm for nourishment. Dwarf males are found in many marine organisms—fish, barnacles, and worms for example—and are especially common in the deep sea, where females can be hard to find. Some dwarf males were originally described as different species or not even recognized as separate organisms at all. In the green spoon worm, a larva on its own turns into a female, but a larva that lands on a female becomes a male. The male is then swallowed and spends its entire life inside the female's uterus.

The most famous sexual parasites are male anglerfish. Each male (there may be a harem here too) is little more than testes held on with a special set of teeth.

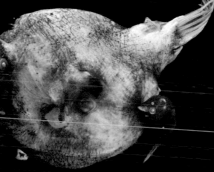

▲ In some pairings, the female has no choice. This large female anglerfish *(Linophryne indica)* plays hostess and mate to the dwarf male attached near her tail.

FAST FACT:
When the sexual stakes are large, as they usually are, males will go to almost any length to sire offspring, literally. Barnacles on the seashore are glued to the rocks, and the nearest female may be some distance away, so the penis of a male barnacle can be eight times as long as the body.

◄ Two male elephant seals *(Mirounga leonina)* fight for dominance of a beach and a harem. In this bloody battle, winner takes all.

HAVING IT BOTH WAYS

▲ For these hamlet fish *(Hypoplectrus unicolor)*, the question is "To be or not to be (female)?" as they take turns spawning eggs and sperm.

FAST FACT:
The Atlantic slipper shell lives in stacks, with females on the bottom and males on top (hence its Latin name, *Crepidula fornicata*). These snails start off male, but if they are alone, they quickly become female. Later arrivals stay male until they graduate to near-bottom status. Paternity tests sometimes reveal that a father has become a mother.

O N LAND, BABIES ARE USUALLY BORN AS ONE SEX AND stay that way. But in the ocean, many fish, shrimps, worms, snails, corals, and other creatures have the option to change sex or be both sexes at once.

Many sex changers start as females and turn into males, some start as males and turn into females, and a few go back and forth. The strategy used depends on what leads to the greatest success. In the blue-headed wrasse, juveniles develop into either males (so-called primary males) or females, but later on, big females can become secondary males. The brightly colored secondary males defend mating territories, and on a good day, a successful male can spawn with as many as 150 females. But the small primary males, which look like females, can also be successful, by sneaking up on spawning pairs and adding their sperm to the mix. Because only big fish can defend a territory, females wait till they are large to switch sexes.

Anemonefish like Nemo switch sexes in the opposite direction. A single male and female share their anemone with bunches of juveniles. The biggest fish is a female, and if she dies, the male becomes a female and a juvenile takes his place. This works out well for both members of the pair because larger females can produce more eggs, whereas larger males have no special advantage since they never have more than one mate.

Being both sexes can be advantageous in deep-sea dwellers like the bizarre tripod fish, where encounters between members of the same species can be rare. More intriguing still are the hamlets, common coral reef fish that are both male and female but spawn in pairs. Because eggs are more work to produce, each fish must ensure that its partner plays fair, and so every evening they take turns being male and female, trading places (and eggs) several times before calling it quits for the night. Not all hermaphrodites are this civilized—some flatworms duel with their penises for the privilege of being the male.

▶ Two blue secondary males swim among a group of yellow females and primary males in this school of blue-headed wrasses *(Thalassoma bifasciatum).*

ALL TOGETHER NOW

FOR FISH THAT SWIM IN SCHOOLS, PRIVACY IS NOT USU-ally an option. To the contrary, aggregating for sex seems to be preferred. Herring mate each fall off the coast of Maine on Georges Bank, gathering together as it gets dark and then spreading out again at dawn. The largest group seen to date was a giant shoal consisting of one quarter of a *billion* fish, weighing in total about 50,000 tons, and stretching as far as 25 miles (nearly twice the length of Manhattan).

But even normally solitary fish, like groupers and snappers, become social when it comes to sex. In the Caribbean, Nassau groupers congregate at traditional gathering sites to reproduce from December through March, swimming as far as 150 miles to mingle with potential mates for about ten days around the full moon. The fish are quite loyal to their rendezvous locales, each individual returning to the same one year after year. What makes these sites so special remains somewhat a mystery, although they are often in places where currents should take the larvae away from waiting predators. Records for some sites go back 100 years, but the fish themselves usually live for less than 20, so knowledge of where to go must be culturally transmitted by experienced fish to the next generation.

Unfortunately, mating hot spots are hard to keep secret. They are well known by local fishers because of the huge catches that are possible when so many fish gather in one place. People are much harder to satiate than other predators, and many traditional spawning sites once used by Nassau groupers are now gone. The numbers of fish at the remaining sites are hugely depleted. In Belize, there used to be ten spawning sites, and now there are only three, and instead of attracting 30,000 fish each, they attract 5,000 or fewer. Once the memory of a site is lost, it seems never to return. So it is no surprise that this once common fish is now listed as an endangered species and only rarely seen. Group sex can be hazardous when people are around.

FEATURED: *Herring, Nassau Groupers, Grunion*

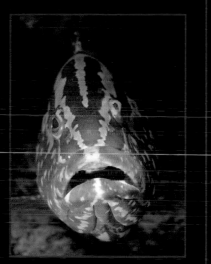

▲ This Nassau grouper *(Epinephelus striatus)* enjoys a solitary lifestyle until the mating season arrives, when it will travel to a traditional group spawning spot.

FAST FACT:
At the highest high tides of spring and summer, the beaches of California and Baja are often crawling with grunion. Female fish surf the waves as far up the shore as they can, where they dig themselves in and spawn. Males then cluster at the surface around the females and release sperm. Two weeks later, the babies are carried back to sea.

◀ In waters near Vancouver, BC, herring *(Clupea pallasii)* swim together in large schools, and when the time comes to reproduce, they spawn together as well.

SPLIT-SECOND TIMING

▶ During the annual mass spawning event, this stony coral *(Acropora species)* releases floating bundles of eggs and sperm into the waters of the Great Barrier Reef.

FAST FACT:
When it comes to who mates with whom in the coral world, timing and location are everything. Paternity-maternity tests (with free-floating offspring, both are needed) show that the mother and father of a baby coral are often neighbors that have spawned within a few minutes of each other.

NINETY MINUTES AFTER SUNSET AND FIVE DAYS AFTER a September full moon, the platy boulder corals of the Caribbean are covered with acne-like pink protrusions. These bundles of eggs glued together by sperm are soon released, first by one colony, then another. The bundles slowly rise to the surface, fertilization occurs, and a new batch of baby corals is launched. After a few nights of this, sex for the platy boulder corals is over for the year.

Scientists still do not know how a simple animal like a coral, without eyes or a brain, manages to tell the time of the day, phase of the moon, and month of the year so precisely—some combination of sunlight, moonlight, and temperature are probably involved. On Australia's Great Barrier Reef, hundreds of species may spawn in a single week, and slicks of eggs floating on the sea surface can be seen from the air and smelled from the shore. Closely related species often differ slightly in spawning times. For example, the closest relative of the platy boulder coral also spawns on the same nights, but its time slot begins about 90 minutes later. Different spawning times help keep eggs from being fertilized by the wrong sperm.

A half hour of sex once a year may seem odd, but synchrony is crucial for organisms that can't move around. Sperm are good swimmers, but they don't survive long, so colonies that synchronize their spawning with nearby colonies of their species are more likely to have their eggs fertilized. And when many species spawn on the same nights (corals are also joined by other creatures), there are so many eggs in the water that the local predators get full before they can eat them all.

Some seaweeds on reefs have their own style of synchronized sex. Like corals, many green seaweeds spawn at precise times, but near dawn rather than at dusk, and different clusters of individuals spawn throughout the year. Unlike their coral brethren, the algae do not survive to reproduce another day. After sex, only clusters of white corpses remain—they have given their all.

▶ These green seaweeds *(Caulerpa racemosa)* on a shallow-water reef off the coast of Panama spawn in groups around sunrise and then die.

TILL DEATH DO US PART

FEATURED: Wandering Albatross, Venus's Flower-Baskets, Shrimps

MONOGAMY IS THE EXCEPTION RATHER THAN THE rule, both on land and in the sea. Even in species where males and females seem to be faithful to their mates, paternity tests often tell another story. For example, although around 90 percent of bird species seem to be monogamous for at least one breeding season, in less than 25 percent of the species are mates actually always faithful. One of the reasons that monogamy is so rare is that it has to benefit both members of the pair. Otherwise, divorce or cuckoldry prevail. Not surprisingly, long-term monogamy is even rarer.

Seabirds are the most monogamous of all birds because raising the offspring almost always requires two committed parents—one member of the pair often guards the offspring while the other heads to sea to find food. The wandering albatross, renowned for its enormous wingspan of as much as 12 feet, is an extreme example. Breeding begins at about ten years of age, but it takes two to three years for a male and female to move from courtship to an actual egg. After the chick hatches, the parents take turns bringing food to the chick for close to a year, from distances that sometimes exceed 500 miles. They then take a year's break from parenting. In a 19-year study of more than 400 pairs, 78 percent bred with their mates, 21 percent were widowed, and the "divorce" rate was less than one percent. Old males without mates sometimes forced paired females to mate, but that was the extent of infidelity.

Sometimes monogamy is enforced by housing arrangements. Several species of shrimps live only inside deep-sea sponges known as Venus's flower-baskets. These sponges are built of a meshwork of glass fibers so fine that Europeans at first thought they were made by Chinese artisans. When young, the shrimps can come and go, but eventually they grow too large to escape, and there is room for only two adults. The beautiful dried sponge skeleton with its entombed pair of shrimps is a traditional wedding present in Japan—it is called

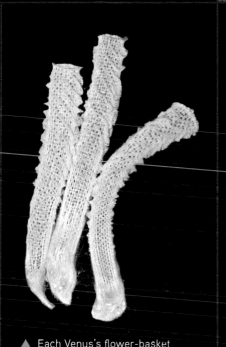

▲ Each Venus's flower-basket sponge *(Euplectella aspergillum)* is home to a male and female shrimp trapped forever in holy matrimony.

FAST FACT:
Many wandering albatross are killed by longlines when they swallow baited hooks. Females are more often victims, which means that unmarried females are in short supply. Young females also prefer males their own age, usually ignoring elderly suitors. This makes it especially hard for the older widowers to find another mate.

DEVOTED DADS

▲ This male sea spider *(Parapallene australiensis)* carries his babies on his back until they are ready to leave and eventually have sea spiders of their own.

FAST FACT:
Both male and female pipefish prefer big mates, though in the end they have to take what they can get. But the offspring of females stuck with small males don't necessarily suffer. Before turning her eggs over to the male, the female broad-nosed pipefish gives them an extra shot of protein if the male is substandard.

I F YOUNGSTERS GET CARED FOR AT ALL, THE MOTHER IS usually involved. But in fish and a few other groups where eggs are not abandoned, fathers are often the primary care providers. Males are sometimes such devoted dads that it takes longer for them to care for the young than it does for the females to produce the eggs. If potential fathers are in limited supply, stereotypical male and female roles get reversed, with males more interested in food and females more interested in sex.

Some male fish build nests. A male will accept the eggs of more than one female, because he only has to defend them and keep them clean. In fact, females often prefer to lay their eggs in nests that already have eggs. There are limits, however—peacock blenny males will reject female offers if the cost of caring for so many eggs gets overwhelming. And male fathers are not entirely trustworthy, as they often eat some of the eggs.

Cardinalfish go one step further—a male carries the eggs of a female in his mouth until they hatch. Here the temptation to eat them is even greater. Small clutches of eggs are especially likely to wind up in the stomach of the father, because forgoing dinner for a week or two is worth it only if lots of babies result. What's a suspicious mom to do? In at least one species, females produce some cheap yolkless eggs that they mix with the normal ones to fool the male into thinking he has a big clutch worth caring for.

Some ocean fathers even have specialized body parts to carry developing eggs. Sea spiders usually use their legs, whereas seahorses and pipefish have a patch or womblike pouch on the belly or under the tail. In both groups, elaborate courtship dances may precede the transfer of eggs from female to male. Seahorses and pipefish have more to worry about than their eggs, however—every year 20 million of the creatures wind up in aquariums or, even more sad, dried as curios or ground up for traditional medicine.

▶ That's quite a mouthful! This male yellow-striped cardinalfish *(Apogon cyanosoma)* forgoes dinner to protect his unhatched offspring.

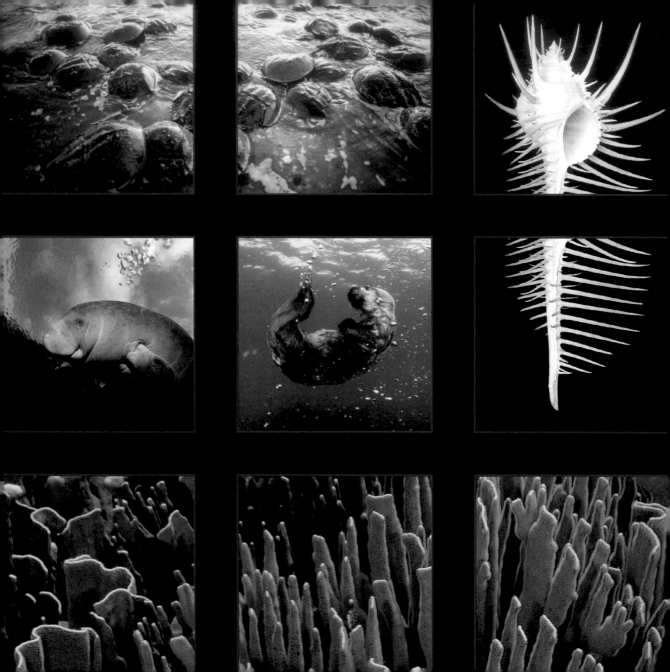

DEEP TIME

Life on Earth began in the sea almost four billion years ago. The steady evolution of new life-forms and the extinction of others have been punctuated at times by moments when nearly everything changed all at once. Yet some creatures look just like their ancestors, and others that were thought long extinct have been "discovered" alive and well. Even land creatures have occasionally returned to their ocean roots.

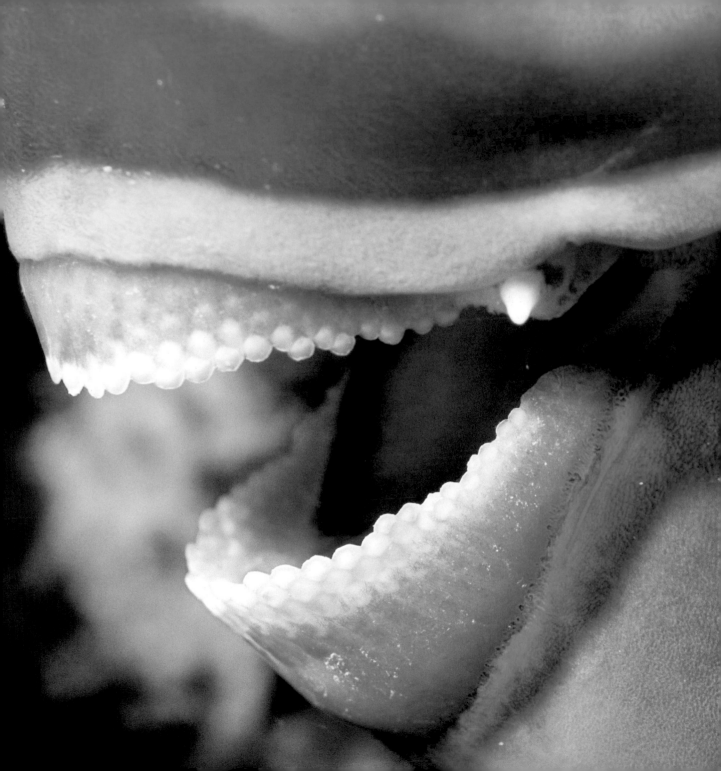

ARMS RACES

G IVEN HUNDREDS OF MILLIONS OF YEARS, PREDATORS in the sea have evolved new ways to capture prey, and their prey have in turn invented new ways to avoid being eaten. This history of escalation has resulted in ever more effective weapons and defenses. The process never stops, but during the age of the dinosaurs, the ocean became a decidedly more dangerous place.

On the predator side, crabs with crushing claws appeared on the scene, and carnivorous snails developed the ability to bore holes through shells. Fish with powerful jaws and strong teeth became experts at breaking open armor. In response, some snails and clams grew thicker and spinier shells. Others cemented themselves to the bottom, making it difficult for their adversaries to get a grip. Still others buried themselves, with just snorkel-like tubes connecting them to the water, with its supply of food and oxygen.

Many groups that did not evolve fast enough went extinct. But some, like the stalked crinoids—relatives of sea stars that filter water for a living—survived. They were once abundant in shallow water but retreated to the safer haven of the deep sea, where there is generally less hustle and bustle. Other survivors escaped the evolutionary turmoil in shallow water caves.

But perhaps the greatest example of the power of arms races occurred about 540 million years ago. Fossils tell the story of an enormous proliferation of different forms of life—in fact, almost every group that we recognize today had its origins during this Cambrian Explosion, named for the geologic era in which it occurred. Suddenly all sorts of different kinds of animals were protected by shells, leaving behind many more fossils than the soft-bodied animals that dominated the seafloor before. The evolution and spread of predators—and the need to defend against them—is the most likely reason that so many animals suddenly found it advantageous to spend the energy needed to build and carry a shell.

▲ The long, sharp spines of this tropical sea snail (*Murex pecten*) deter predators that can't handle its prickly defense.

FAST FACT:
Though the deep sea may provide a partial haven for some species, it's hardly conflict free. One snail from a hydrothermal vent has a triple-plated defense that is unusually strong and crack resistant, probably to protect against crushing crab claws. Here an animal arms race may contribute to the human equivalent—the development of better armor.

◄ "Smiling" for the camera, this colorful ember parrotfish (*Scarus rubroviolaceus*) shows off its impressive chops, which can easily pulverize coral.

FROZEN IN TIME

FAST FACT:
Horseshoe crabs (and some other sea creatures) have blue blood because a protein with copper, rather than iron, is used to trap oxygen. Their blood is colorless when oxygen is scarce, so if a horseshoe crab "turns blue," that's a good sign. Some worms have yet another pigment that turns green if oxygen runs low.

THE GREEK PHILOSOPHER HERACLITUS BELIEVED THAT nothing is permanent but change itself. The fossil record is in general a testament to this, with its continuous parade of innovations punctuated by mass extinctions. Two hundred fifty-one million years ago, 95 percent of the species in the sea died out, and another large chunk of diversity vanished 65 million years ago with the meteorite that also wiped out the dinosaurs. But a few groups seem to have persisted through this turmoil without changing much at all.

Horseshoe crabs are the most famous of these "living fossils" in the sea. The body plan and ecology of these relatives of scorpions have been roughly the same for at least 200 million and perhaps as many as 360 million years. Horseshoe crabs live on sandy and muddy bottoms, feeding on worms and clams. Beachcombers occasionally find their shells, and in late spring and early summer they come into shallow water to mate and lay their eggs on the beach. Migrating shorebirds like red knots depend on these eggs to fuel their flights.

Why does evolution seem to grind to a halt for some groups of organisms? Often these groups have just a few species—there are only four species of horseshoe crabs, and one species of blue coral, another living fossil. Most evolutionary change occurs when new species are formed, so that in groups where new species rarely form, there are fewer opportunities for novelty to arise. Of course, most of these sluggish lineages die out, but a few, like the horseshoe crabs, survive.

Having persisted through the ages, today horseshoe crabs are on the decline from destruction of their coastal habitat and from human harvesting. Fishers use horseshoe crabs for bait, and their blue blood is also valuable. The clotting factor that it contains is used to detect bacterial toxins—it even was carried to the international space station to test surfaces for unwanted bacteria. It seems the height of injustice that an animal whose family tree dates back hundreds of millions of years could vanish in 50 years thanks to human activity.

▶ Some family trees go way back. Blue corals *(Heliopora coerulea)* are the only surviving species of a lineage that dates back to the time of the dinosaurs.

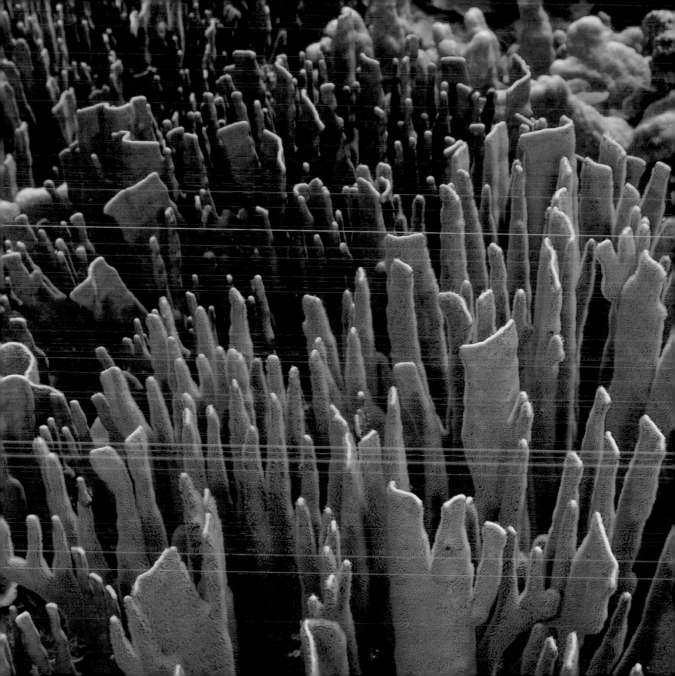

▲ During the high tides of late spring, these horseshoe crabs (*Limulus polyphemus*) lay their eggs on the sandy beaches of New Jersey.

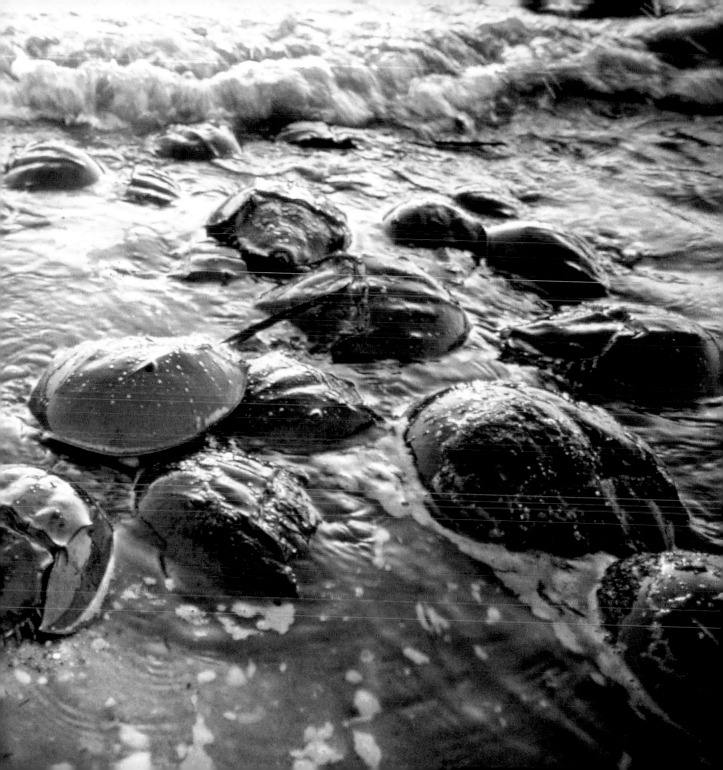

BACK FROM THE DEAD

JESUS IS SAID TO HAVE BROUGHT LAZARUS BACK TO LIFE four days after his death. Biologists claim no miracles, but they do have their "Lazarus species," organisms once thought to be extinct that have been rediscovered alive and well. Sometimes this is good news for conservation. For example, a fire coral from the Galápagos was thought to have died out during the severe El Niño of 1982–83, but it was then rediscovered alive several years later.

The most spectacular revivals are those of species thought to be extinct for many millions of years—living fossils in every sense. Living fossils can tell us things that their dead cousins cannot, since they have soft tissues, behaviors, and DNA that can be studied. For example, the fossils of seemingly obscure, snail-like creatures called monoplacophorans, known from hundreds of millions of years ago, were intriguing because a row of paired markings on their shells suggested they might be related to the eight-shelled chitons or, possibly, segmented worms. The collection of a living monoplacophoran from a deep-sea trench in 1952 caused a biological sensation, and the specimen was even once hidden in the museum so that it couldn't be stolen.

The most spectacular Lazarus species are coelacanths. Thought to have gone extinct before the demise of the dinosaurs, they were discovered alive off the African coast in 1938. These "fish" are actually more closely related to us than they are to tuna or cod. Their two pairs of lobed fins are precursors to the four limbs of land-dwelling frogs, birds, and mammals, and their swimming even resembles walking in the way the fins are moved. An entirely separate species of coelacanth was discovered in a fish market in Indonesia in 1997.

Lazarus species are a vivid reminder of how much of the planet remains unexplored, especially in the deep sea, and what more we might still discover. Their continued survival, however, is hardly guaranteed. Coelacanth numbers are declining, suggesting that they could go extinct once again, but this time for good.

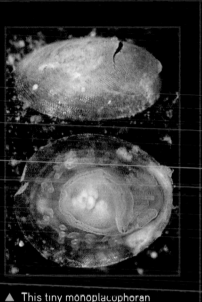

▲ This tiny monoplacophoran *(Laevipilina hyalina)*, shown from above and below, was found clinging to rocks in 1,300 feet of water off the coast of San Diego.

FAST FACT:
A living fossil shrimp, thought to be extinct for 50 million years, was collected in 1908. Since then, only a few dozen have been found, one most recently in about 1,500 feet and dubbed Jurassic shrimp by Census scientists. In the deep sea, knowing where to look helps—good collecting locations for living monoplacophorans have recently been found in California.

◄ Until it was captured by fishers, the coelacanth (*Latimeria* sp.) was thought to have gone extinct more than 65 million years ago.

▲ Though highly poisonous, most sea snakes, like this banded sea krait *(Laticauda colubrina)* from Indonesia, generally only bite when provoked.

FAST FACT:
For snakes that have gone back to the sea, dehydration is a problem. They get some freshwater from their food, but most need to drink. For a few sea snakes, the ocean is not an option—even with salt glands, it's "water, water, everywhere, nor any drop to drink."

ALTHOUGH LIFE FIRST AROSE IN THE SEA, MOST OF THE animals with backbones that we are familiar with, except for fish, evolved on land. The majority are still firmly rooted there, but some have made the return trip—sea turtles, sea snakes, marine iguanas, seabirds, and marine mammals.

Marine mammals are arguably the most successful invaders in terms of the number of species and the degree to which they have left the land behind. Three important groups have gone back to ocean life: the whales and dolphins; the walruses, seals and sea lions; and the dugongs and manatees. As individual species, sea otters and polar bears have also made the transition.

The most recent invaders are sea otters and polar bears. Polar bears are close relatives of brown bears, and probably originated only about one million years ago, during the ice ages. Sea otters are related to freshwater otters, and more distantly to weasels and mink, and they date back less than five million years. Seals, sea lions, and walruses have an older ocean pedigree. Their nearest relatives are probably bears, and their aquatic lifestyle began about 30 million years ago.

Manatees and whales have had the longest time to adapt to the sea—more than 50 million years. The manatees and dugongs are the only plant-eating marine mammals, and their closest relatives, the elephants, are also plant eaters. Whales and dolphins have a vegetarian ancestry too, having evolved from hippo-like animals, but they are now carnivores.

Changes are needed for a once terrestrial animal to succeed at sea. Water is a heat sink, so fur or fat is used for insulation. Limbs and tails are modified for swimming or lost altogether. Fossils reveal the history of these changes and provide missing links, like the whales with tiny back legs found in the desert sands of the Middle East. After 300 million years, why would all these groups go back to the sea? In every case, the reason seems to be to get more food.

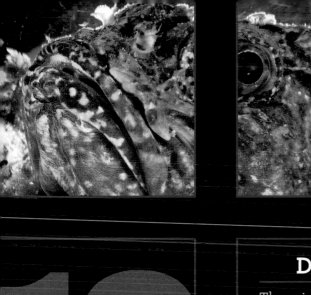
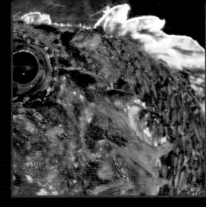

DANGEROUS ENCOUNTERS

12

There is an understandable fascination with the ocean creatures that can kill us. Some of them bite, some sting, some poison us only if we eat them. They range in size from giant crocodiles to microscopic bacteria. Usually, it is simply a case of our being in the wrong place at the wrong time, as very few marine organisms make a living by doing us harm. We pose a far greater threat to life in the sea than sea life poses to us.

EATEN ALIVE

MANY COASTAL PEOPLES ARE TRADITIONALLY AFRAID of the ocean and often cannot swim. These fears are not irrational—the oceans were once so teeming with predators that swimming would have been tantamount to suicide. Early explorers of Pacific atolls had to fight off sharks with their oars when they tried to come ashore. It is only because we have killed most of the top predators that the oceans today seem so tame by comparison—shark populations are currently a tiny fraction of what they used to be, and some species have declined by 75 percent in just the past 20 years. But even so, we fear being attacked.

Sharks are the most notorious of ocean man-eaters, although most sharks in fact are harmless. Of the approximately 400 species of sharks, only about 30 have been reported as attacking people, and only a dozen represent a serious threat. The most dangerous are the great white, tiger, and bull sharks. In many cases, attacks on people are probably cases of mistaken identity.

In tropical waters, sharks are not the only large things that eat people. The saltwater crocodile is a fearsome predator that can reach more than 20 feet in length. Although the water is its natural home,

▲ Despite its formidable teeth, humans are more of a threat to the great white shark *(Carcharodon carcharias)* than the great white is to humans.

FAST FACT:
The wide-ranging lifestyle of sharks, their slow reproductive rate, the high value of their fins, and the fact that

SEA STINGERS

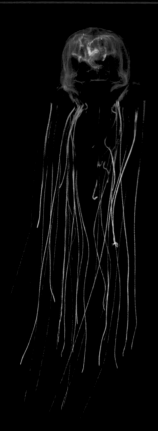

▲ In Australia, nets protect swimmers against this large species of sea wasp *(Chironex fleckeri)*, but the smaller sea wasps can get through.

FAST FACT:
Fireworms defend themselves with a poison-laced battery of microspines that they puff out at the slightest disturbance. These sword swallowers of the sea first open wide and engulf coral branches, then back off after digesting the living tissue.

JUST A LIGHT BRUSH AGAINST A FIRE SPONGE, FIRE-worm, or fire coral can result in a sharp burning sensation. Scorpionfish and stingrays deliver their more powerful punch with venom-laced spines. In some cases, being stung is simply unpleasant, but it can result in excruciating pain or even death.

The most infamous lethal stingers are the sea wasps, or box jellies. Box jellies look superficially like regular jellyfish, but they are not closely related. They have a cubical design (hence their name) and one or more tentacles hang from each of the four bottom corners. The tentacles are covered with thousands of stinging cells, each equipped with a microharpoon and poison sac. They also have 24 eyes and, unlike regular jellyfish, are agile swimmers. The most feared is *Chironex fleckeri*, which in Australia can be common in shallow water all summer long. Their danger is because of not just the toxicity of the venom, but also the amount—the swimming bell of these giants can be nearly a foot in diameter, with tentacles that extend for as far as ten feet. When a person is wrapped in their stinging embrace, death can occur in minutes from cardiac arrest. But perhaps even scarier, because they can pass through the nets that keep the big box jellies away from beaches, are those that are no larger than a peanut, with tentacles the width of only a hair. Unlike their more notorious cousins, the initial reaction to the sting is much less severe, and no welts are left behind. But within an hour the symptoms of what is called Irukandji syndrome develop—back pain, nausea, cramps, sweating, rapid pulse, and a feeling of impending doom. This extreme anxiety is sometimes justified, as death from cerebral hemorrhage due to skyrocketing blood pressure has occurred more than once.

Poisons are ubiquitous in nature because they are so useful. For some creatures, like sea wasps, they are used to subdue their prey. For others, like stingrays, they keep potential predators at bay. For us, they are a potent reminder to treat sea life with the respect it deserves.

▶ This scorpionfish *(Scorpaena papillosa)* can be hard to spot, but grabbing it or stepping on it by mistake makes for an unforgettably unpleasant surprise.

▲ A stingray *(Dasyatis americana)* can inflict a painful stab when stepped on by a careless wader. The culprit is the poisonous spine nestled near the base of the stingray's tail.

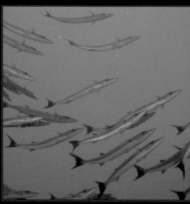

CULINARY DISASTERS

FEATURED: *Fugu, Blue-Ringed Octopuses, Dinoflagellates, Barracuda*

NORMALLY, GOING TO A SEAFOOD RESTAURANT IS NOT considered a death-defying experience. But Japanese diners will pay as much as $400 for fugu, a fish so toxic that it contains enough poison to kill 30 people. The especially poisonous skin, liver, intestines, and gonads must be carefully removed, and fugu chefs take a course whose final exam consists of preparing and (more important) eating the delicacy. These days, most deaths from fugu poisoning are caused by fugu do-it-yourselfers.

Death from fugu poisoning is a gruesome affair—paralysis shuts down the ability to breathe, but the victim remains completely mentally aware. What makes fugu, a kind of pufferfish, or blowfish, so poisonous is tetrodotoxin, or TTX, which is hundreds of times more deadly than cyanide. The poison is made not by the fish but by the marine bacteria in its food. Fugu are not alone in having TTX—the blue-ringed octopus, notorious for its lethal bite, cultivates TTX-producing bacteria in its salivary gland. Haitian voodoo doctors are said to use pufferfish in the powder that turns people into zombies.

Bacteria are not the only poison producers in the sea. Dinoflagellates (dinos, for short) are single-celled organisms, and some of them are highly toxic and cause the things that eat them to be toxic too. When some dinos bloom and form red tides, the clams and oysters that filter them out of the water can cause paralytic shellfish poisoning when eaten. Ciguatera poisoning is caused when another type of dino is eaten by small fish, which are eaten by big fish and then eaten by us. At each step in the food chain the toxins get concentrated (as happens with man-made poisons, such as mercury), so that big fish-eaters like barracuda are the most dangerous. Ciguatera is the most common type of seafood poisoning worldwide. Though rarely fatal, symptoms can be very unpleasant, including the bizarre sensations of cold things feeling hot and hot things feeling cold.

Some seafood can be downright hazardous to your health.

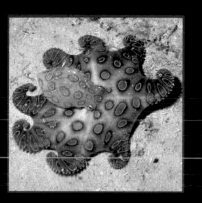

▲ With the same poison as the deadly fugu fish, this blue-ringed octopus *(Hapalochlaena maculosa)* can deliver a lethal bite.

FAST FACT:
It's now possible to raise poison-free fugu by carefully controlling what they eat, making fugu-style foie gras safe for the first time. But fugu without fear is not universally welcomed—it may put some highly trained chefs out of business, and a hint of danger is clearly part of the appeal.

◀ So delicious, they're to die for! Be careful when pufferfish *(Takifugu rubripes,* above) or barracuda *(Sphyraena* sp., below) are on the menu.

SEA SICK

▲ This harmful algal bloom, a "red tide," stretched for more than 20 miles as it approached the coast of southern California.

FAST FACT:
Harmful blooms of small planktonic organisms not only poison seafood but also produce aerosols that make people sick, even those who stick to meat and potatoes. Being on the beach during a red tide can result in eye irritation, coughing, runny noses, and asthma attacks—not exactly the desired outcome of a seaside vacation.

I N 1831–32, PEOPLE WERE DYING IN DROVES IN LONDON, Paris, and New York. The cause was cholera, a disease that, if untreated, can kill more than half of its victims. It is caused by a comma-shaped bacterium that produces a toxin that destroys the inner lining of the intestines. The result is diarrhea so severe that sufferers can die of dehydration within 24 hours, one of the most rapidly fatal of all human diseases. The 11th President of the United States, James Polk, was killed by cholera, as was the composer Tchaikovsky. Fortunately, cholera can be treated by giving the patient large amounts of sugary or salty water, to keep blood watery and in chemical balance.

Although the cholera bacterium, *Vibrio cholerae,* was discovered in 1854, it was not until 1977 that scientists realized that it was common in coastal waters around the world, from the Bay of Bengal to the Chesapeake. In all these places, *Vibrio cholerae* hitches a ride with one of the most abundant animals in the plankton—the copepod—which is why cholera epidemics start on the coast and then move inland. Just one copepod can carry 10,000 of the deadly bacteria.

Although cholera is no longer a serious problem in developed countries, it continues to kill in Africa, Asia, and Latin America—more than 200,000 deaths since 1950. Prevention is always the best medicine, so knowing that cholera and copepods are inextricably linked has given public health officials new tools. It's not possible to get rid of copepods, or for that matter desirable—they are the primary food of many of the fish that we eat. But satellite images that monitor the condition of the ocean can give a heads-up when conditions especially favorable to copepods appear. And because copepods are a lot bigger than bacteria, it is much easier to clean the water by simple filtration. In India, where cholera likely originated and where it has killed tens of millions of people, just pouring water through the fine silk that saris are made of does the trick.

▶ Though this copepod (*Paraeuchaeta* sp.) hails from Antarctica, its tropical relatives can transport cholera to waters all around the world.

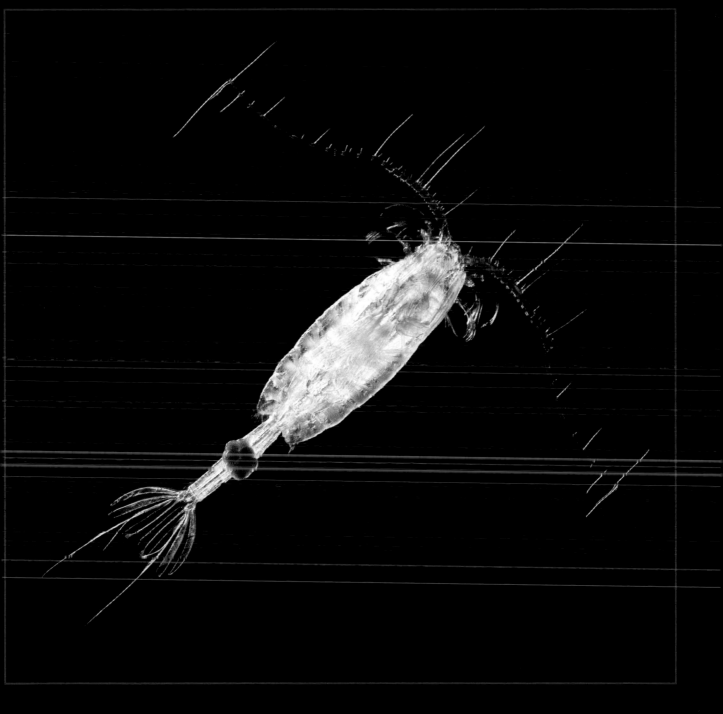

13

FOR WHAT IT'S WORTH

The sea has been generous to humans, and we have profited from this generosity. Bluefin tuna and pearls are worth a king's ransom, and we have scoured the sea for them. Less expensive ocean products are found in everything from ice cream to toothpaste. Ocean creatures have provided new drugs for pain and cancer, and new ideas for industry. But the sea's most valuable gifts are those that we often don't pay for.

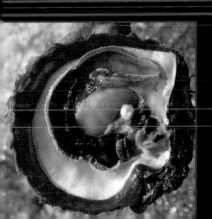

MEGABUCKS

THERE IS PLENTY OF MONEY TO BE MADE FROM SEA CREA-tures. Ocean jewels in the form of pearls, black corals, and red corals fetch hefty prices. Unlike gemstones, ocean jewels are potentially renewable resources, but collectors have often decimated natural populations of the animals that create these exquisite materials. After the Spaniards arrived in the Americas, pearl oysters quickly vanished from the scene, and red corals could soon be protected under the Convention on International Trade of Endangered Species (CITES).

Pearls may be a girl's second-best friend, but in Japan, bluefin tunas are often worth more. They are so highly prized for sashimi that single fish are auctioned off for tens or even hundreds of thousands of dollars. The laws of supply and demand mean that as the fish get rarer, the prices go up and the pressure to catch them increases, creating a vicious circle. Not surprisingly, bluefin tunas are also potential candidates for protection by CITES.

Conservation options are limited when prices are sky-high. Physically protecting such valuable species at sea is almost impossible—patrolling for poachers is extremely difficult, and reaching agreements to control legal fishing is hard when so much money is at stake. Changing consumer preferences can help reduce the demand side of the equation by making it unfashionable to eat or wear endangered species. But the best solution is figuring out how to grow these animals sustainably.

This has worked well for pearls. Oysters feed by filtering small bits of food from the water, so once they get big enough, they can take care of themselves—they just need a safe perch to grow on. And pearls last a long time. But tunas are another matter. Not only are they consumed, but "farmed" tuna is really a misnomer. Small tunas are caught and placed in pens, and because they are predators, it takes 20 or more pounds of fish to produce a pound of tuna. When we figure out how to grow our tunas from eggs and raise them as vegetarians, then the money made from tuna will be a lot greener.

▲ The nondescript shell of this pearl oyster (*Pinctada* sp.) hides a valuable cultured pearl. Now mostly farmed, pearl oysters are very rare in the wild.

FAST FACT:
Adult Atlantic bluefin tunas spawn in the Gulf of Mexico and the Mediterranean, and like salmon, most go back to where they were born to reproduce. But the chemistry of their ear stones shows that some European teenage tunas hang out along the shores of the United States, complicating the management of this endangered fish.

◀ Worth nearly its weight in gold, the critically endangered bluefin tuna *(Thunnus*

FEATURED: Seaweeds

▲ It may not look that delicious, but the carrageenans from "Irish moss" seaweed *(Chondrus crispus)* help keep ice cream creamy.

FAST FACT:
Seaweeds come in three varieties—red, brown, and green—that aren't actually related to each other at all. Green seaweeds, though less used by industry, have a special claim to fame. They invaded the land more than 400 million years ago and gave us the trees and flowers that surround us today.

FANS OF JAPANESE FOOD ARE WELL AWARE OF THE culinary potential of seaweeds, and many other Asian cuisines use seaweeds extensively. Monosodium glutamate (MSG), though now manufactured from scratch, originally was what made dried kelp a flavor enhancer in China. But even if you never go near a sushi restaurant, you are likely to be eating or otherwise using seaweeds.

Red seaweeds provide carrageenans, agar, and agarose, while brown seaweeds provide algins and alginates—these substances are nearly ubiquitous in modern life because they thicken and stabilize creams, sauces, pastes, and other liquids. In ice cream, carrageenans keep ice crystals from forming. In chocolate milk, they keep the particles of cocoa from settling to the bottom. The head of your beer is made possible by alginates. Onion rings and other breaded foods contain them, as do frostings and salad dressings. In tinned hams, agar keeps the meat from sticking to the can.

It is not just in food that seaweeds reign. You probably use their products every morning if you soften your skin with lotion or brush your teeth with toothpaste (and if you wear dentures, algins were probably used to make them too). Scientific laboratories around the world are dependent on agar and agarose— agar fills petri dishes in which bacteria are grown, and agarose gels help biochemists separate bits of DNA. Seaweed extracts are mixed into paint and used to make paper. Kelps provide key ingredients of organic fertilizers. Seaweeds were even once used to stiffen silk, helping garments retain their form after washing.

Because of the huge demand for seaweeds, they are now grown around the world—more than eight million tons in the year 2000 alone. In some cases, they have escaped from the farms, causing problems. But many things we take for granted are made possible by the gels and goos that seaweeds provide. Rumor even has it that carrageenan had a starring role in the fake saliva of the monster in *Alien!*

▶ Near the British Isles, a diver swims among free-living kelps *(Laminaria hyperborea).* Elsewhere, in underwater farms, kelps are grown in bulk for industry.

NEPTUNE'S PHARMACY

THE HEALTH BENEFITS OF EATING FISH ARE WIDELY promoted. And if you are sick, the sea has drugs to offer. Most are made by animals, plants, fungi, and bacteria to kill their prey or fend off their competitors. Though these toxins are deadly in large doses, in the right amount many of them can have important medical applications. Zorivax, from a sponge, is used against herpes infections. Some extracts from sea slugs, sea squirts, and moss animals kill cells or stop them from reproducing, making them potentially important in the fight against cancer. One example is Ecteinascidin, derived from a sea squirt and sold under the name of Yondelis.

Cone snails are masters of chemical warfare. Often exquisitely beautiful, these animals are deadly predators. Their teeth are specialized venom filled harpoons that paralyze and kill their prey. Some cones are so deadly that they have killed people who innocently picked them up. Cones are hugely diverse, with about 700 species, and each of these produces a unique cocktail of 100 to 200 toxins. Some cone toxins are extraordinary painkillers—ziconotide, or Prialt, which has been on the market since 2004, is more than 100 times more powerful than morphine. Others may help with epilepsy or heart attacks. Just a handful of cone toxins have been clinically tested, so we have barely scratched the surface of their medical potential.

Finding drugs in the sea is not as easy as it sounds. Because the compounds are so toxic, organisms usually produce them in small amounts, and just getting enough to study is difficult. Trials that look promising in a test tube often fail in humans, because either the drugs don't work as expected or the side effects are too severe. If the drugs do work, figuring out how to produce them in commercially useful amounts is the next challenge. But in the case of cone shells, one big

▲ This cone snail *(Conus dalli)* from the Sea of Cortés may one day follow in the footsteps of its cousin the magician cone and provide new painkillers.

FAST FACT:
In the past 30 years, more than 15,000 chemicals have been discovered in ocean organisms. Because our traditional medicines have not often come from the sea, we have less experience to guide us in our search for new cures. Sometimes, it's the microbes in organisms rather than the organisms themselves that are the drug producers.

INSPIRED DESIGN

FEATURED: Humpback Whales, Sharks, Boxfish, Mussels

▶ The ridged flipper of this breaching humpback whale *(Megaptera novaeangliae)* has inspired new designs for wind turbines.

FAST FACT:
A newspaper article extolling the benefits of new types of tapes (including one inspired by the feet of geckos) noted that tape failure is usually caused by people failing to dry surfaces—"if it's wet, nothing will stick to it except mussels in the ocean." Can mussel tape be far behind?

THREE BILLION YEARS HAS GIVEN MOTHER NATURE A LOT of time to be creative, so the solutions to many of our problems have already been created. Biomimetics is the science of turning nature's designs to our advantage. Velcro, for example, was inspired by the annoying ability of burs to cling to our clothes.

Animals that move through the ocean have come up with many ingenious ways to deal with water flow, and we have copied some of them. The tiny teeth in shark skin have inspired everything from high-tech swimsuit materials to antifouling surfaces on ships. Mercedes-Benz drew on the shape of a boxfish to create a car with low drag and lots of storage space. The bumpy design of the fins of the humpback whale, when used for wind turbine blades, makes it possible to generate the same amount of power with 60 percent less wind.

Not all ocean solutions are about efficient movement—other organisms survive by holding on. Mussels and seaweeds are champions at this, gluing themselves onto wave-swept surfaces. The glues they use are nontoxic, work in water, and even adhere to nonstick cookware. But getting just a few pounds of mussel glue would require more than five million mussels, so imitating the structure of these glues is a lot more practical. Glues inspired by a California worm that makes bits of sand and shell stick together could both repair bone fractures and deliver pain medicine and antibiotics at the same time. Studies of corals have already led to the invention of an injectable paste that can repair damaged ends of arm and leg bones. Someday, the metal screws and plates now used by orthopedic surgeons will be largely obsolete (and their biomimetic replacements will have the added benefit of not setting off metal detector alarms).

The power of natural selection rewards good designs and punishes bad ones, so both incentives and quality control are as much a part of the natural world as they are a part of the commercial world. We simply need to look for the sea's good ideas.

▶ Sometimes necessity is the mother of invention, and sometimes nature is. Mussels, boxfish, and the scaly skin of sharks are all sources of ideas for industry.

ECONOMISTS ARE GOOD AT FIGURING OUT WHAT SOMEthing is worth if it can be owned, bought, and sold. But there is no market for some of what the sea provides—oysters and sponges filter and clean our water "for free," making it trickier to figure out their value.

What, for example, is a sea otter worth? Dead, they used to be worth a lot, and that was how their value was calculated—their fur was so sought after that they nearly went extinct before being protected in 1911. To calculate their value today, economists add up incomes and jobs from otter-watching, plus the amount people say they are willing to pay just to know that sea otters exist, plus the otters' value in keeping kelp forests healthy, minus the value of the seafood they eat that we might otherwise harvest. One such calculation came up with a price tag of millions of dollars for several hundred California sea otters. But this kind of analysis is not easy to do well, which is why nature is so often taken for granted.

Undervaluing nature leads to bad decisions. The mangrove forests that line tropical shores are routinely destroyed to make room for hotels and farms. The economic benefits of hotels and farms are easy to calculate, but the costs associated with cutting down mangroves are not, so this side of the equation tends to be ignored. As a result, more than one-third of all mangroves have disappeared in just the past 20 years, a more serious loss than that suffered by tropical rain forests. The economic consequences are considerable, even if they are hard to measure. In the Gulf of California, nearly a third of all commercially valuable fish and crabs depend on mangroves, making the true worth of mangroves more than $15,000 per acre, far more than the government estimates. And in some places mangroves do far more than provide food. In India, a super typhoon in 1999 killed nearly 10,000 people, but death rates were lower in villages buffered by a wide band of mangroves. Protecting nature literally saved lives.

◀ With an abalone (Haliotis sp.) in hand, this sea otter (Enhydra lutris) returns to the surface of the Pacific Ocean to feast and entertain tourists.

▲ Mangrove forests provide nurseries for fish, protect shorelines from waves, and could absorb carbon dioxide, making them more valuable alive than dead.

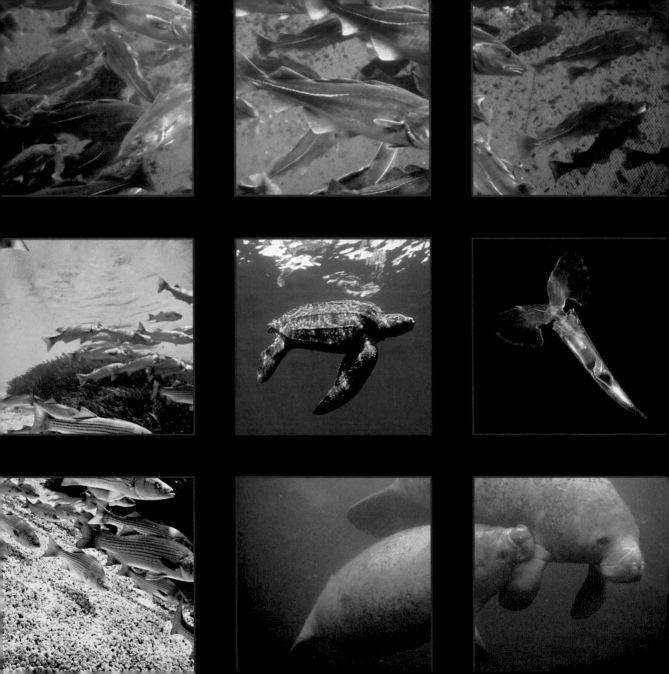

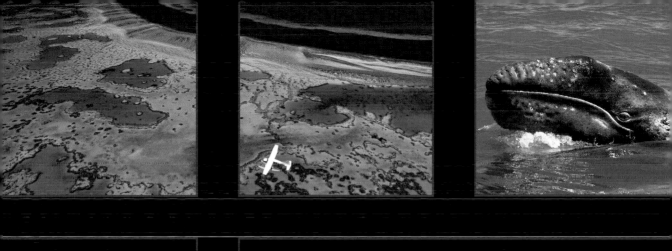

14

BAD NEWS, GOOD NEWS

Despite the ocean's grand scale, humans have managed to change it profoundly, causing some ocean species to go extinct and pushing many more to the edge of existence. Ecosystems have been transformed by fishing, pollution, invasions, and most recently, global warming and changing ocean chemistry. But if we work together, conservation and restoration are possible. Indeed, signs of recovery can already be found.

NINETEENTH-CENTURY SCIENTISTS BELIEVED THAT ocean organisms were too numerous and wide-ranging for humans to drive to extinction. And indeed, to date just three mammals, five birds, two fish, two snails, and two seaweeds are thought to have vanished in historic times as a result of human activity.

The mammals and birds were largely victims of human hunters using fairly primitive methods—most were extinct and all were rare by 1800. The sea mink was gone before it could be described by scientists (only bones and teeth remain), and the great auk was never seen by scientists in the wild (the last pair was taken in 1844). The number of Steller's sea cows on two arctic islands was never greater than 3,000, and in 1768 the last ones were killed, only 17 years after their discovery. These giant relatives of dugongs and manatees reached nearly 25 feet in length and five tons in weight. Just one could keep 33 men eating for a month, but they were hunted at a rate seven times what could be sustained. Among extinct mammals, the longest survivors were the Caribbean monk seals, which hung on in small numbers until the 1950s. By then they were also probably starving because of humans fishing their food. These historic extinctions represent the tail end of a wave of losses started by prehistoric human hunters as they moved across the planet, eliminating mastodons, moas, and other megafauna.

The other, smaller creatures known to have vanished at our hands were species with limited habitats, such as a single harbor, which were made uninhabitable by human activities. But the true number of such extinctions is probably much higher—species that have literally disappeared without a trace. Many more species still alive today could already be doomed to extinction just from lack of space.

They say extinction is forever, which is why we fight to prevent it. Future generations may be able to resurrect a few lost creatures, *Jurassic Park* style—this idea has inspired efforts to create a frozen Noah's ark. But most extinct species are gone for good.

▲ Because the great auk *(Pinguinus impennis)* went extinct before the camera went mainstream, we have only illustrations and stuffed birds to remind us how it looked.

FAST FACT:
The loss of Steller's sea cows seems especially tragic. Confronted by their human killers, they were reported to show an "uncommon love" for one another, with some of them attempting to remove harpoons from their wounded comrades. They were the only cold water and seaweed-eating sea cows, and their four remaining tropical relatives, including manatees, are all at risk.

◀ A close relative of the extinct Caribbean monk seal *(Monachus tropicalis),* the Hawaiian monk seal *(Monachus schauinslandi)* is also highly endangered.

AT THE BRINK

▶ Relatives of the extinct Steller's sea cow, this mother and child manatee *(Trichechus manatus)* and their Pacific cousins the dugongs are also in trouble.

FAST FACT:
Leatherback turtles need protection not only from poachers at their nesting beaches but also from fishing boats at sea. In the eastern Pacific, tagged turtles follow predictable pathways that suggest where protection is most needed. Their odysseys of thousands of miles have inspired online enthusiasts who follow the Great Turtle Race, organized by Census scientists.

FEW SPECIES MAY HAVE VANISHED ENTIRELY, BUT MANY are at risk. One-third of all corals are considered in danger, making them the most threatened group of animals on the planet. And some species are truly at the brink, pushed to the edge of existence by fishing, habitat destruction, or the changing climate.

Big animals often suffer most from human activities, but the smallest porpoise is the most endangered. The vaquita (Spanish for little cow) lives only at the northern tip of the Gulf of California. Every year, dozens are trapped and drowned in shrimp trawls and fishing nets set for sharks and the totoaba fish (which are also endangered). Vaquita numbers have declined by more than two-thirds since 1997 and now fewer than 200 remain. The solution—removal of all fishing nets from where vaquita survive—is clear. But those nets support Mexican fishing villages, and addressing the human side of the conservation equation has proven to be very difficult. With its distinctive black lip and eye markings, the vaquita is one of the most beautiful of porpoises, but its shy habits make it an unlikely source of income from tourists. At current rates of killing, it will be gone in just a few years.

Although the vaquita's situation is unusually dire, many other species also hang on by a thread. White abalone numbers are so low that most males and females are too far apart for sperm to reach eggs. Nesting populations of Pacific leatherback turtles have declined precipitously. Once numbers get very low, they can plummet further for other reasons, such as bad weather—only a few colonies of an already rare Galápagos fire coral survived the warm waters brought by the devastating 1982–83 El Niño, and hundreds of Florida manatees died in the severe winter of 2009–2010. Offspring of rare species may also have genetic problems due to inbreeding. Once at the brink, there is literally no room for error, and moving survivors into captivity, even if feasible, carries its own risks. Avoiding the crisis is obviously far better than dealing with it after the fact.

COSTLY COLLAPSE

WHEN MARINE POPULATIONS COLLAPSE, OCEAN dwellers are not the only victims. People depend on the ocean for food and jobs, and when management fails, they suffer the consequences. Unfortunately, collapsing fisheries are all too common—one in four fisheries has collapsed in the past 50 years.

The northern cod of Canada is the poster child for a costly collapse. These once abundant fish fed Europe and the Americas for more than 500 years. The seas off the coast of Newfoundland were described in 1497 as "swarming" with fish, so numerous that they could be caught by throwing a weighted basket over the side of a ship. The famous *bacalao* of Portugal and salt fish of Jamaica were cod. But greed essentially, consuming nature's capital rather than living off the interest— brought this once profitable and proud fishery to its knees. Today, the amount of northern cod is less than 4 percent of what it was in the middle of the 19th century.

When the fishery was closed in 1992, it resulted in the biggest loss of jobs Canada has ever seen, and it has since cost the government about three billion dollars. Unfortunately, the damage done by unsustainable fishing can be long lasting—the cod have not come back. The reasons for this failure to recover are not entirely clear. It might be because other fisheries continue to catch baby cod, or because the sea is getting warmer. Or the entire food web may have been thrown out of whack—the fish that adult cod used to eat are now much more common, and they may be eating most of the few baby cod that are produced. As is often the case, a combination of causes is probably at work.

The saga of the northern cod was a sad and costly lesson—ocean resources are not infinite, and biological and economic catastrophes occur when we take more from the sea than the sea can sustain. Yet we continue to make many of the mistakes that led to the cod's collapse— the cod's lesson seems a hard one to learn.

FAST FACT:
Captain John Smith was so impressed by the seemingly limitless number of fish off the coast of New England that his 1635 map showed a vast school between Cape Cod (note the name) and Cape Ann in Massachusetts. Dried cod was the fourth most valuable export from the American Colonies just before the Revolution.

◀ Though it may seem that arctic cod *(Gadus morhua)* are flourishing in fish farms like this one off Newfoundland, wild populations have been decimated by overfishing.

▲ Without any predators to control them, invasive "killer algae" *(Caulerpa taxifolia)* crowd out other species in the Mediterranean Sea.

FAST FACT:
San Francisco Bay may be the most invaded body of water in the world. In the second half of the 20th century, one new species invaded San Francisco Bay every 14 weeks, and by 2000, the bay had 157 non-natives. Dumping ballast water at sea rather than in ports helps slow the rate of introductions.

NOT ALL OCEAN TRAGEDIES ARE ABOUT COLLAPSING populations—the problem can be too much of something rather than too little. Remove top predators or add nutrients, and the oceanic balance of nature is upset. Jellyfish now dominate some coastal seas, causing one Japanese fishing boat to capsize due to their weight in the nets. Takeovers by species that were introduced, by mistake or on purpose, can have especially disastrous results.

The Mediterranean has been transformed by the invasion of more than 500 species, particularly two related species of Australian "killer algae." Initially spreading at rates of more than 200 miles per year, these seaweeds are now found in locations stretching from Spain to Cyprus. Mediterranean seaweed eaters avoid them because of the nasty chemicals they contain, allowing the invaders to monopolize space at the expense of more diverse and productive native communities.

Pacific lionfish had never been seen in the Atlantic until they started turning up near Florida in 1985. By the year 2000, lionfish were seen all along the East Coast of the United States and in Bermuda. By 2004, they were in the Bahamas, and since 2007 they have spread rapidly through the Caribbean. Densities in the Bahamas are now greater than 150 per acre. These highly efficient predators can reduce the number of baby fish on a patch of reef by nearly 80 percent in just five weeks, bad news for a region that is already catastrophically overfished.

When it comes to invasive species, prevention is best, but that can be hard given the many possible entryways—from ships' hulls and ballast water, aquaculture, canals, and aquariums (the source of both lionfish and killer algae). The next best solution is a rapid response like that which eliminated killer seaweeds in California—an effort that cost more than seven million U.S. dollars. Once an invasive species is established, control (even more expensive) is the only option left. In the case of the lionfish, restaurants that serve it and earn some money in the process help make the best of a very bad situation.

▶ First spotted off the East Coast of the United States, this non-native lionfish *(Pterois volitans)* has now spread throughout the Caribbean.

DISSOLVING AWAY

THE INDUSTRIAL REVOLUTION HAS TRANSFORMED NOT only the lives of people but also the life of the planet. Increasing carbon dioxide in the atmosphere from burning oil, coal, and natural gas has caused and is causing the Earth (and ocean) to get warmer. But about one-third of the carbon dioxide released has not warmed the planet because it has dissolved into the ocean. Although this sounds like a good thing, it comes at a price—dissolved carbon dioxide makes the ocean more acidic.

Why does acidity matter? The more acid the ocean, the harder it is to lay down the stony material that makes up the skeletons and shells of many ocean organisms. Some corals lose their skeletons entirely if the water gets too acidic. For many reef dwellers, it is the skeleton that creates the nooks and crannies that they call home, so a coral without a skeleton is not much better than no coral at all.

But homes are not the only things at stake—these changes in ocean chemistry can make food scarcer as well. Many small fish depend on tiny shelled snails, called pteropods, for food—they are such an important source of energy that they have been called the potato chips of the sea. Smaller fish feed bigger fish and bigger fish feed us, so we too are dependent on the tiny pteropods. Shellfish such as oysters, scallops, clams, and mussels also have to spend extra energy to lay down their shells in an acid ocean, so acid oceans are bad for aquaculture as well.

The global warming side of the carbon dioxide story has gotten most of the media attention—dying corals, monster storms, and rising sea levels are a product of hotter seas. But the prospect of an ever more acidic ocean actually worries many ocean scientists more. Temperatures have risen and fallen often during the history of the Earth, but acidity is usually much more stable. No one knows what the outcome of this giant chemistry experiment will be, but there is little reason to hope that osteoporosis of the ocean will be a good thing.

▼ The delicate shell of the half-inch-long pteropod snail (*Creseis virgula*) will easily dissolve in the acid waters of the future.

FAST FACT:
Stony skeletons differ in the chemical details of their construction, and these differences affect their vulnerability to the effects of acid oceans. The seaweeds that grow as stony pink crusts are especially sensitive. This is bad news not only for them but also for corals, whose babies like to settle on them.

◀ Today, a hermit crab (*Paguritta harmsi*) pokes its head out of its coral home, but as seas become acid it may join the ranks of the homeless.

▲ A collection of old menus from the 1950s serves as proof that things aren't what they used to be when it comes to seafood.

FAST FACT:
In colonial America, lobsters were not the several-pound creatures of today but 20-pound giants. Orphans and prisoners begrudgingly ate them, as they had no choice, but servants insisted in their contracts that they not be fed lobster more than twice a week. Only in the 1800s did tastes start to change.

WHAT DOES A HEALTHY OCEAN LOOK LIKE? WITH each passing generation, the definition of what is normal is reset—what our grandparents took for granted, we find unimaginable. This phenomenon of shifting baselines means that we often accept as healthy what is actually sick. Defining "healthy" for the ocean is a challenge because scientists didn't begin studying the ocean until relatively recently. To figure out what the ocean really looked like before people had an impact requires some detective work.

Samples collected by paleontologists and archaeologists can tell us quite a lot. Fossil reefs from thousands of years ago show that the degraded reefs of today have no precedent. Mud from decades and centuries past, collected by hammering pipes into estuaries like the Chesapeake Bay, tells a story of more nutrients and less oxygen as people increasingly dominated the landscape. By rummaging through the garbage piles of prehistoric peoples, we know that North American Indians used to dine on oysters the size of dinner plates.

People have long kept detailed records of what interested them. Explorers and pirates wrote about what they saw—stories of beating back sharks with oars in places where sharks no longer can be found are vivid testimony to how the world has changed. Bookkeepers kept careful track of economically important species, allowing us to back-calculate how many green turtles and cod there used to be.

Even old menus, postcards, and souvenir photographs provide a surprising window into the ocean of the past. Lobsters used to be used as fertilizer and were eaten only by the poor. An abalone dinner that today costs upwards of $80 was $6.50 in the 1920s, a price rise ten times as great as the rate of inflation. The average trophy fish in Key West declined from 44 to 5 pounds between 1956 and 2007. Though shocking, these pictures of what we have lost are also inspirational. They show us what we could return to, and they are worth far more than a thousand words.

▶ Once valuable on a dinner plate, this goliath grouper *(Epinephelus itajara)* could rake in more dough in front of the lens of a camera.

ON THE ROAD TO RECOVERY

THE SEEMINGLY ENDLESS STORIES OF DOOM AND GLOOM can make even the most optimistic of ocean lovers despair. But as we have recognized the gravity of the situation and taken action, things have started to turn around for some ocean dwellers.

Green turtles are one such source of hope. These huge vegetarians used to be so abundant that sea captains spoke of being able to navigate by their sounds. In the Caribbean, they numbered about 90 million and easily outweighed the large animals of the Serengeti. Not surprisingly, the English and Spanish turned to green turtle flesh and eggs to feed their colonies, and it did not take long for turtle numbers to be decimated (hence the invention of mock turtle soup). But thanks to protection, their numbers have lately been increasing by anywhere from 4 to 14 percent a year (a healthy interest rate) at five beaches where many females come to lay their eggs. Though other former nesting beaches still have few or no turtles, the trends in at least a few places are positive. Some whales have also been increasing in number since the end of most commercial hunting in 1986. Protection of large charismatic species can work.

But this species-by-species approach clearly has its limits—there are simply too many kinds of organisms, many of which are less glamorous than whales and turtles, to protect them all individually. So, for most ocean dwellers, it is far more effective to protect the real estate where they live, just as we do on land. The gold standard is Australia's Great Barrier Reef, where one-third of the park is completely protected from fishing. Other megaparks have also been established or are being planned. In these large parks, numbers and sizes of fish have rebounded, and outbreaks of disease and pest species have declined.

Glimmers of hope need to be recognized for what they are—a fragile start to the process of rebuilding. Time and again, fisheries have been reopened with the first signs of recovery, only to crash anew. Nevertheless, celebration is in order for what good news we do have.

▲ Green sea turtles *(Chelonia mydas)*, like this one from the Red Sea, still need protection from egg poachers to ensure numbers continue to increase.

FAST FACT:
The North Atlantic right whales were hunted to near extinction, but even after the end of whaling, whales continued to die from being hit by ships. Recent rerouting of ship tracks and lower speed limits seem to be helping—2009 saw the birth of a record 39 calves, practically a baby boom.

BACK IN BUSINESS

▲ A gray whale *(Eschrichtius robustus)* breaches in a lagoon in Baja California, where it has come to breed.

FAST FACT:
Eastern Pacific gray whales do more than just entertain tourists in whale-watching boats. Because they suck up lots of mud (and animals) from the sea bottom and bring it to the surface, they also provide food to seabirds. On average, every gray whale helps feed 11 seabirds in the Bering Sea.

EVER LONGER LISTS OF ENDANGERED AND THREATENED species are a sign of failure—the goal is to take species off these lists (and not through extinction). Though listed species still greatly outnumber delisted ones, there are some genuine comebacks.

One such saved species is the striped bass, one of the first fish to be managed in the United States. In 1649 Massachusetts made it illegal to use stripers as fertilizer because of their importance for food, and in 1670 the first American public school was partially paid for by taxes on their sale. But simply requiring that stripers only be eaten was, not surprisingly, not enough. By the mid-1700s, concerns about imprudent and even wanton fishing were being raised, and in the 1970s, catches plummeted.

Stripers were vulnerable on a number of fronts, making it easy for everyone to point the finger at someone else. Because they spawn in estuaries, they suffer from the effects of pollution and dammed rivers, and their late maturity (eight years) makes them sensitive to overfishing. In the 1980s, laws were finally passed to protect them, and today harvests have rebounded, much to the joy of the recreational fishers who catch most of them. Another animal now off the official worry list is the eastern Pacific gray whale. In this case the business model has changed, as the money to be made comes from whale-watching, not whaling.

Back in business does not mean home free. Given the size of the human footprint, nothing really is—ocean organisms can no longer simply take care of themselves. Many stripers are now diseased or in poor condition, and polychlorinated biphenyls (PCBs) in their flesh make eating them dangerous for young children and pregnant or nursing women. Eastern Pacific gray whales, even at 22,000 strong, are still probably well below their original numbers. There will always be problems. But if we take our role as ocean (and climate) guardians seriously, we can look forward to a future when we can again enjoy the bounties of the ocean with pleasure and without guilt. For this, the citizens of the sea will thank us.

▶ Thanks to careful management, numbers of striped bass *(Morone saxatilis)* have rebounded, to the delight of fishers.

ABOUT THE CENSUS

MORE THAN 2,000 SCIENTISTS. 80-PLUS NATIONS. 400 expeditions. $650 million. 10,000-plus possible new species. 2,500-plus publications. And 22 million database records and counting.

These numbers only begin to describe the ten-year international effort to study life in the global ocean undertaken by the Census of Marine Life in 2000. It set out to assess the diversity (how many different kinds), distribution (where animals live), and abundance (how many) of marine life in the global ocean—a task never attempted before. Not only did the first Census of Marine Life prove that such a sweeping study could be done, it also demonstrated how much still remains to be learned.

The first Census of Marine Life created a baseline picture of ocean life—past and present—that can be used to forecast, measure, and understand changes in the marine environment as well as to inform the management and conservation of marine resources. It accomplished this by investigating life in the global ocean, from microbes to whales, from top to bottom, from Pole to Pole. This required international collaboration on an unprecedented scale, which brought together the world's eminent marine biologists, who shared ideas, data, and results. It also required a major commitment to technology that would provide a window into the world below the waves. Census scientists took full advantage of available technology and significantly added to that toolbox with innovative techniques and new adaptations and creations, along with standardized protocols that will well serve the next generation of scientists.

Over a decade of discovery, Census scientists uncovered many of the ocean's long-held secrets. They found species previously thought to be extinct and identified about 200 new marine species, with tens of thousands more waiting for formal designation as such. They followed tuna as they moved across the Atlantic and back, first in search of food, then in search of mates. They identified areas in the ocean where animals congregate from white shark cafés to deep-sea vents where animals live without light and feast on chemicals from the Earth's core. Scientists documented long-term and widespread declines

in marine life and areas where recovery was apparent. They also better identified what is unknown, what needs to be known, and what may remain unknowable about life in the global ocean.

In the end, the Census has laid a foundation for the direction of future research and science-based policy and served as a model for how such huge scientific undertakings can be structured for success. Most important, perhaps, is that the Census was able to demonstrate the connection between life in all ocean realms and all ocean basins, capturing the wonder, beauty, and magic of life in the global ocean along the way.

—Ian Poiner, *chair,*
Scientific Steering Committee,
Census of Marine Life

CENSUS
OF MARINE LIFE

The following individuals served as Census of Marine Life liaisons to the National Geographic Society and as advisers for this book:

Sara Hickox
Education and Outreach Team Leader

Patrick N. Halpin
Mapping and Visualization Team Leader

Darlene Trew Crist
Communications Team Leader

For more information about the Census of Marine Life, including images, reports, and a complete list of Census projects, please visit *www.coml.org.*

ABOUT THE AUTHOR

NANCY KNOWLTON SPENT HER CHILDHOOD SUMmers on the shores of Long Island Sound and was educated at Smith College, Harvard University, and the University of California, Berkeley. She learned to dive in 1972, inspired by working for the first woman to descend to the deep-seafloor in the submersible *Alvin*. Her research on coral reefs has taken her to Jamaica, Panama, Brazil, and remote atolls of the central Pacific and Indian Oceans. After holding positions at Yale University and the Smithsonian Tropical Research Institute, she founded the Center for Marine Biodiversity and Conservation at the Scripps Institution of Oceanography of the University of California, San Diego. Now the Sant Chair for Marine Science at the Smithsonian's National Museum of Natural History and a scientific leader of the Census of Marine Life, Knowlton has devoted her life to studying, celebrating, and striving to protect the multitude of life-forms that call the sea home. She lives with her family in Washington, D.C.

ACKNOWLEDGMENTS

WITHOUT AMANDA FEUERSTEIN THIS BOOK would not have happened. She researched stories, found images, wrote captions, suggested titles, and was relaxed and cheerful no matter how urgent the deadline or long the hours.

The Census of Marine Life was the inspiration for this book and the source of much of its content. I owe a special thank-you to Darlene Trew Crist for her unflagging efforts to get me stories and photographs; to Frank Baker for his help with the Census's photographic archive; and to Jesse Ausubel for making the Census happen and supporting this celebration of it.

The publishing team at National Geographic transformed my ideas into a beautiful reality. I thank Kevin Eans and Melissa Farris for their aesthetic vision; Marshall Kiker and Garrett Brown for editing and patiently shepherding the project, and Enric Sala for writing the foreword.

The staff of the Smithsonian helped in many ways, especially the library staff, Christine Hoekenga, and Yolanda Riley, as well as the many people who were understanding when I couldn't do anything except finish the book.

My scientific colleagues at the Census, the Smithsonian, the Scripps Institution of Oceanography, and elsewhere helped me tell their stories with ideas, images, and information. Thank you, Greg Rouse (who provided material for several stories), Maria Baker, Bodil Bluhm, Ann Bucklin, Gisella Caccone, Roy Caldwell, John Christy, Ken Clifton, Allen Collins, Mireille Consalvey, Nancy Copley, Emmett Duffy, Casey Dunn, Brigitte Ebbe, Peter Franks, Kim Fulton-Bennett, Melissa Garren, Rolf Gradinger, Michael Hadfield, Roger Hanlon, Hiroki Hata, David Helvarg, Ana Hilario, Russ Hopcroft, Jennifer Jacquet, Michael Lang, Don Levitan, Eva Ramirez Llodra, Loren McClenachan, Chris Meyer, Alvaro Migotto, Don Moore, Jon Norenburg, Karen Osborn, Shirley Pomponi, Jez Roff, Joe Sacco, Scott Shaffer, Ellen Strong, Jonathan Thar, Edward Vanden Berghe, Michael Vecchione, Gritta Veit-Kohler, Robert Vrijenhoek, Janie Wulff, Jeremy Young, and others I have forgotten.

I thank Roger and Vicki Sant, who have championed the ocean and supported me and my work at the Smithsonian from the beginning. Thank you also to the many government and private funders who have supported the research that informs these pages.

Finally, my family. Far more than words can express, thanks to my mother and father for encouraging me to be a scientist; to Sally and Frank for sharing the wisdom of schoolteachers; to Stephen and Rebecca, who are the most supportive, kind, and patient children one could ask for; to my mother-in-law Faith Jackson, whose publishing a book at the age of 90 made me believe I could do it at 60; and to my biggest supporter and marine scientist extraordinaire, my husband, Jeremy Jackson.

INDEX

All of the images on the endsheets and chapter openers are repeated in the interior except where noted. 1, Susan Middleton; 2-3, Paul Sutherland/NationalGeographicStock.com; 7, Gary Cranitch/Queensland Museum; 8, Norbert Wu/Minden Pictures/NationalGeographicStock.com; 11 (UP RT), Mike Veitch/SeaPics.com; 12, David Wrobel/SeaPics.com; 13, Photo Ifremer/A. Fifis; 14, "On some frenulate species (Annelida: Polychaeta: Siboglinidae) from mud volcanoes in the Gulf of Cadiz (NE Atlantic)," Ana Hilário and Marina R. Cunha – 2008, Sci. Mar. 72(2): 361-371; 15, Bill Curtsinger/NationalGeographicStock.com; 16 (UP), Doug Perrine/SeaPics.com; 16 (LO), Tanya Burnett/SeaPics.com; 18-19, Flip Nicklen/Minden Pictures/NationalGeographicStock.com; 20, Carola Espinoza; 21, Flip Nicklin; 22, Melissa Garren; 23, Jeremy Young/Natural History Museum, London; 24, Christopher Meyer; 25, Chris Newbert/Minden Pictures/NationalGeographicStock.com; 28, Fred Bavendam/Minden Pictures/NationalGeographicStock.com; 29 (LO), Chris Newbert/Minden Pictures/NationalGeographicStock.com; 29 (UP), Flip Nicklen/Minden Pictures/NationalGeographicStock.com; 30-31, Paul Nicklen; 32, Paul Nicklen/NationalGeographicStock.com; 33 (UP), Birgitte Wilms/Minden Pictures/NationalGeographicStock.com; 33 (LO), Roger Hanlon; 34, Brian Skerry; 35, Karen Osborn; 36, Gary Cranitch/Queensland Museum; 37, Darlyne A. Murawski/NationalGeographicStock.com; 38, (c) 2004 MBARI; 39, Peter Parks/SeaPics.com; 41 (UP RT), Joel Sartore; 42, Paul Zahl; 43, Debi Henshaw; 44, George Grall/NationalGeographicStock.com; 45, George Grall/NationalGeographicStock.com; 46, Gary Cranitch/Queensland Museum; 47, Randy Olson; 48, Norbert Wu/Minden Pictures/NationalGeographicStock.com; 49, William Van Orden/BilzRockFish/COML; 50, Fred Bavendam/Minden Pictures/NationalGeographicStock.com; 51 (UP), Marc Chamberlain/SeaPics.com; 51 (LO LE), Chris Newbert/Minden Pictures/NationalGeographicStock.com; 51 (LO CTR), Mark Jones/Minden Pictures/NationalGeographicStock.com; 51 (LO RT), Fred Bavendam/Minden Pictures/NationalGeographicStock.com; 52, Mark Jones/Minden Pictures/NationalGeographicStock.com; 53, Robert F. Sisson; 55 (UP RT), Brian Skerry; 56, Flip Nicklen/NationalGeographicStock.com; 57, Paul Nicklen/NationalGeographicStock.com; 58, David B. Fleetham/SeaPics.com; 59 (UP), Wolcott Henry/NationalGeographicStock.com; 59 (LO), Joel Sartore; 60, Stefano Unterthiner; 61, Paul Nicklen; 62-63, Alexander Safonov; 64, Russ Hopcroft/

COML; 65, Chris Gotschalk/Wikipedia; 66, Doug Perrine/SeaPics.com; 67, Phillip Colla/SeaPics.com; 68, Jez Roff; 69, Larry Madin; 71 (UP RT), Peter Parks/SeaPics.com; 72, Fred Bavendam/Minden Pictures/NationalGeographicStock.com; 73, Cheryl Clarke Cmarz/UAF; 74-75, Robert F. Sisson; 76, Albert Lleal/Minden Pictures/NationalGeographicStock.com; 77, (c) 2006 MBARI/NOAA; 79 (UP), Espen Rekdal/SeaPics.com; 79 (LO LE), Phillip Colla/SeaPics.com; 79 (LO CTR), Wolcott Henry/NationalGeographicStock.com; 79 (LO RT), NOAA; 80, New Zealand's National Institute of Water and Atmospheric Research; 81, Franco Banfi/SeaPics.com; 83 (UP RT), Emory Kristof; 84, Mona Lisa Production/Photo Researchers, Inc.; 85, Stephen Low Productions, Inc.; 86, Russ Hopcroft; 87, Kevin Raskoff; 89 (UP), COML; 89 (LO LE), Michael Aw/SeaPics.com; 89 (LO CTR), Buntzow/Corgosinho; 89 (LO RT), Kevin Raskoff; 90, David B. Fleetham/SeaPics.com; 91, Norbert Wu/Minden Pictures/NationalGeographicStock.com; 92, Wikipedia; 93, Carol Buchanan/SeaPics.com; 95 (UP RT), Tui De Roy/Minden Pictures/NationalGeographicStock.com; 96, Uwe Piatkowski; 97, Solvin Zankl/SeaPics.com; 98-99, David Doubilet; 100, Gilbert S. Grant/Photo Researchers, Inc.; 101, Hiroki Hata; 102, Tui De Roy/Minden Pictures/NationalGeographicStock.com; 103, Tui De Roy/Minden Pictures/NationalGeographicStock.com; 105 (UP), Mauricio Handler/NationalGeographicStock.com; 105 (LO), (c) 2003 MBARI; 106, Emory Kristof/NationalGeographicStock.com; 107 (UP RT), D. R. Schricte/SeaPics.com; 110, Mark Strickland/SeaPics.com; 111, Doug Perrine/SeaPics.com; 112-113, Doug Perrine/SeaPics.com; 114, Doug Perrine/SeaPics.com; 115, Jeremy Stafford-Deitsch/SeaPics.com; 116, Mary S. Tyler; 117, Carol Buchanan/SeaPics.com; 118, Janie Wulff; 119, Dr. Jonathan Dowell, ReefNews.com; 121 (UP RT), David B. Fleetham/SeaPics.com; 122, Hiroya Minakuchi/Minden Pictures/NationalGeographicStock.com; 123, Hiroya Minakuchi/Minden Pictures/NationalGeographicStock.com; 125, David Doubilet; 126 (UP), Doug Perrine/SeaPics.com; 126 (LO), Mark Conlin/SeaPics.com; 128, Emmett Duffy; 129, Emmett Duffy; 131 (UP RT), Ken Clifton; 132, Alvaro Migotto; 133, Tui De Roy/Minden Pictures/NationalGeographicStock.com; 134, John Eastcott & Yva Momatiuk; 135, Norbert Wu/Minden Pictures/NationalGeographicStock.com; 136, Steven Kovacs/SeaPics.com; 137, NOAA CCMA Biogeography Team; 138, Paul Nicklen; 139, Norbert Wu/Minden Pictures/NationalGeographicStock.com; 141 (UP), Fred Bavendam/Minden Pictures/NationalGeographicStock.com; 141 (LO), Ken

Clifton; 142, Paul Nicklen/NationalGeographicStock.com; 143, Susan Dabritz/SeaPics.com; 144, John C. Lewis/SeaPics.com; 145, Chris Newbert/Minden Pictures/NationalGeographicStock.com; 147 (UP RT), Emory Kristof; 148, Paul Sutherland/NationalGeographicStock.com; 149, Wikipedia; 151, Andre Seale/SeaPics.com; 152-153, Fred Bavendam/Minden Pictures/NationalGeographicStock.com; 154, Dr. Hans W. Fricke; 155, Greg Rouse; 156, Rodger Klein/SeaPics.com; 157 (UP), Paul Nicklen; 157 (LO LE), Bill Curtsinger/NationalGeographicStock.com; 157 (LO CTR), Paul Nicklen/NationalGeographicStock.com; 157 (LO RT), Brian Skerry/NationalGeographicStock.com; 159 (UP RT), Dr. Anthony R. Picciolo, NOAA NODC; 160, Franco Banfi/SeaPics.com; 161, Mike Parry/Minden Pictures/NationalGeographicStock.com; 162, David Doubilet; 163, Brian Skerry; 164-165, James D. Watt/SeaPics.com; 166 (UP), Norbert Wu/Minden Pictures/NationalGeographicStock.com; 166 (LO), Wolcott Henry/NationalGeographicStock.com; 167, Clay Bryce/SeaPics.com; 168, Peter Franks; 169, Ingo Arndt/Minden Pictures/NationalGeographicStock.com; 171 (UP RT), Chris Newbert/NationalGeographicStock.com; 172, Mike Parry/Minden Pictures/NationalGeographicStock.com; 173, David Doubilet; 174, Image courtesy Andrea Ottesen/University of Maryland/Department of Plant Science and Landscape Architecture/Science; 175, Simon Brown/SeaPics.com; 176, Doug Perrine/SeaPics.com; 177, Norbert Wu/Minden Pictures/NationalGeographicStock.com; 179 (UP), Alaska Stock Images/NationalGeographicStock.com; 179 (LO LE), Richard Herrmann/SeaPics.com; 179 (LO CTR), Chris Newbert/Minden Pictures/NationalGeographicStock.com; 179 (LO RT), Doug Perrine/SeaPics.com; 180, Bates Littlehales; 182-183, Tim Laman; 185 (UP RT), Ralph Lee Hopkins/NationalGeographicStock.com; 186, Bill Curstinger/NationalGeographicStock.com; 187, George Bernard/Photo Researchers, Inc.; 189, James A. Sugar; 189 (LO LE), Doug Perrine/SeaPics.com; 189 (LO CTR), Dr. Kevin D. Lafferty, USGS; 189 (LO RT), Flip Nicklen/NationalGeographicStock.com; 190, Doug Allan/SeaPics.com; 192, Reinhard Dirscher/SeaPics.com; 193, Michael Patrick O'Neill/SeaPics.com; 194, Chris Newbert/Minden Pictures/NationalGeographicStock.com; 195, Russ Hopcroft; 196, Glenn Jones; 197, Doug Perrine/SeaPics.com; 198, David Doubilet; 199, Aviram Avigal; 200, Dr. Steven Swartz, NOAA/NMFS/OPR; 201, Doug Stamm/SeaPics.com; endsheet (LO CTR), Ingo Arndt/NationalGeographicStock.com.

CITIZENS OF THE SEA: WONDROUS CREATURES
FROM THE CENSUS OF MARINE LIFE
Nancy Knowlton

PUBLISHED BY THE NATIONAL GEOGRAPHIC SOCIETY
John M. Fahey, Jr., *President and Chief Executive Officer*
Gilbert M. Grosvenor, *Chairman of the Board*
Tim T. Kelly, *President, Global Media Group*
John Q. Griffin, *Executive Vice President; President, Publishing*
Nina D. Hoffman, *Executive Vice President;*
    President, Book Publishing Group

PREPARED BY THE BOOK DIVISION
Barbara Brownell Grogan, *Vice President and Editor in Chief*
Marianne R. Koszorus, *Director of Design*
Lisa Thomas, *Senior Editor*
Carl Mehler, *Director of Maps*
R. Gary Colbert, *Production Director*
Jennifer A. Thornton, *Managing Editor*
Meredith C. Wilcox, *Administrative Director, Illustrations*

STAFF FOR THIS BOOK
Garrett Brown, *Editor*
Marshall Kiker, *Project Manager and Illustrations Specialist*
Melissa Farris, *Art Director*
Kevin Eans, *Illustrations Editor*
Amanda Feuerstein, *Picture Legends Writer*
    and Story and Illustrations Researcher
Lisa A. Walker, *Production Project Manager*
Al Morrow, *Design Assistant*
Allison Gaffney, *Design Intern*

MANUFACTURING AND QUALITY MANAGEMENT
Christopher A. Liedel, *Chief Financial Officer*
Phillip L. Schlosser, *Vice President*
Chris Brown, *Technical Director*
Nicole Elliott, *Manager*
Rachel Faulise, *Manager*

The National Geographic Society is one of the world's largest nonprofit scientific and educational organizations. Founded in 1888 to "increase and diffuse geographic knowledge," the Society works to inspire people to care about the planet. It reaches more than 325 million people worldwide each month through its official journal, *National Geographic,* and other magazines; National Geographic Channel; television documentaries; music; radio; films; books; DVDs; maps; exhibitions; school publishing programs; interactive media; and merchandise. National Geographic has funded more than 9,000 scientific research, conservation and exploration projects and supports an education program combating geographic illiteracy. For more information, visit nationalgeographic.com.

For more information, please call 1-800-NGS LINE
(647-5463) or write to the following address:

National Geographic Society
1145 17th Street N.W.
Washington, D.C. 20036-4688 U.S.A.

Visit us online at www.nationalgeographic.com

For information about special discounts for bulk purchases, please contact National Geographic Books Special Sales: ngspecsales@ngs.org

For rights or permissions inquiries, please contact National Geographic Books Subsidiary Rights: ngbookrights@ngs.org

ISBN: 978-1-4262-0643-6

Library of Congress Cataloging-in-Publication Data

Knowlton, Nancy.
Citizens of the sea : wondrous creatures from the census of marine life / Nancy Knowlton.
    p. cm.
Includes index.
ISBN 978-1-4262-0643-6 (hardcover : alk. paper)
1. Marine animals--Behavior. 2. Marine ecology. I. Title.
    QL121.K56 2010
    591.77--dc22
                     2010012894

Printed in China

15/PPS/5

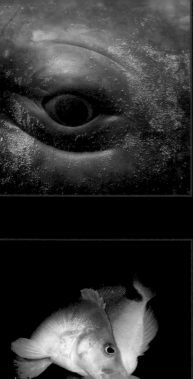

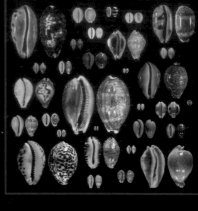

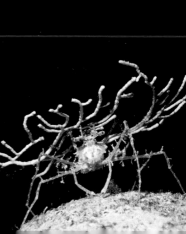